A SHOT IN THE ARM

How Science, Engineering, and Supply Chains
Converged to Vaccinate the World

Yossi Sheffi

MIT CTL MEDIA

Cambridge, Mass.

MIT CTL Media
77 Massachusetts Avenue, Building E40
Cambridge, MA 02139

This book was set in Alkes and Cooper Hewitt by
MIT CTL Media. Produced in the United States of
America. Front cover design by Arthur Grau.

Published 2021
Library of Congress Control Number: 2021948639
ISBN-13: 979-8-9850705-0-7

Library of Congress Cataloging-in-Publication Data

Sheffi, Yossi, 1948–
A shot in the arm: how science, engineering, and
supply chains converged to vaccinate the world / Yossi
Sheffi.
Includes bibliographic references.

Table of Contents

Preface

When the Covid-19 pandemic struck in early 2020, I set aside a project on the history of innovation in supply chains to write a book analyzing what ultimately became perhaps the greatest global supply chain disruption since World War II. That book, *The New (Ab)Normal: Reshaping Business and Supply Chain Strategy Beyond Covid-19* (MIT CTL Media, 2020), focused on how Covid impacted the broader economy, having major disruptive effects on consumer goods, manufacturing, distribution, logistics, retailing, workplaces, and cities. In the same vein as two of my previous books dealing with risk and resilience—*The Resilient Enterprise: Overcoming Vulnerability for Competitive Advantage* (MIT Press, 2005) and the *Power of Resilience: How the Best Companies Manage the Unexpected* (MIT Press, 2015)—*The New (Ab)Normal* looked at how people and companies were handling, adapting to, and even benefiting from disruptive events—both Covid-related and in general.

In *The New (Ab)Normal*, I described how employees were working from home and consumers were battling to hoard toilet paper (detailing the origins of this and other shortages), as well as how companies were navigating the combined impacts of changing consumer demand, disrupted suppliers, fractured transportation links, and new workplace regulations. At the same time, I began watching another story unfolding in the laboratories of universities and pharmaceutical companies. Biomedical scientists and engineers around the world began a race to save civilization from the virus by developing a vaccine. Those scientists and engineers seemed to face very long odds of success in any reasonable timeframe given both the very long gestation periods typically required to create just the candidate vaccines for testing and the low rate of subsequent approvals of tested vaccines.

Many likened this massive vaccine development effort to the Apollo moonshot campaign of the 1960s. However, the moonshot was easier in several ways. The moonshot only needed to hand-build a dozen or so rockets to carry a select few dozen intrepid astronauts willing to risk it all to go into space and then to the moon. In contrast, the vaccine effort needed to mass-produce billions of units of a safe and effective product for use by billions of ordinary consumers and citizens, many of whom were risk-averse and hesitant. Whereas the moonshot's rocket makers could deliver the rockets to a single location, the vaccine makers needed to reach everyone in the world wherever they lived. Finally, while US President John F. Kennedy gave NASA more than eight years to reach the moon, Covid began killing staggering—and increasing—numbers of people from day one.

This new book follows the global efforts to vaccinate the world and begins with a chapter focused on the accumulated knowledge in the sciences of immunology, genetics, and cell biology that made the vaccines possible. Literally decades of science and more than a few Nobel Prizes went into understanding how viruses and vaccines work and creating the tools for making vaccines. Over time, scientists came to a fundamental understanding of how the cells of every living creature are really biological factories, how viruses hijack those cellular factories, and how the immune system learns to fight off the hijackers. One key innovation in the newest generation of vaccines (the mRNA vaccines of Moderna, Pfizer-BioNTech, and others) is in how they make the antigen—the Covid spike protein—that the immune system learns to recognize and fight. The mRNA vaccines don't contain the antigen; rather, they contain a set of instructions, the equivalent of a business purchase order (PO) to a person's cells so that those cells manufacture this most important component of the vaccine. However, those innovative mRNA vaccines could not work without other innovative technologies that package the exceedingly fragile mRNA and deliver it to people's cells.

Developing a safe and effective vaccine wasn't the end of the challenge; it was just a first step in what would become the greatest product launch in human history: mass-producing these vaccines, distributing them to vaccination sites around the world, and getting billions of people to come and get vaccinated—discussed in the following chapters.

The second chapter of this book focuses on the next step of the challenge: mass-producing an entirely new product from scratch and rapidly scaling production to millions—and then billions—of units within the shortest possible time. Mass-producing the vaccine meant creating all the supply chains needed to manufacture all the ingredients and raw materials required for the vaccine, many of which had been niche laboratory chemicals. Getting to scale entailed overcoming shortages of materials and industrial capacity.

After the vaccine makers produced and bottled all the doses, next came the challenge of getting the product to the people who needed it (described in Chapter 3). This is analogous to the challenge of getting the hottest new toy out to retailers' shelves for the holidays, but with the added pressure that impediments and delays in distribution would mean that more people would die. This task of consumer distribution was managed by governments with varying degrees of effectiveness. The mix of happy successes and woeful failures offer many lessons for how to handle new product launches, especially when product demand overwhelms supply.

As with any new product launch, product availability is one half of the distribution challenge with consumers. Chapter 4 examines the other half: how of the vaccine providers had to convince, cajole, or incentivize people to go out and get the product. Although a great many people eagerly wanted the vaccine, detractors of vaccination arose as a serious threat to the prevention of death and disease. These so-called anti-vaxxers were a messy mix of those with heartfelt personal values, the sadly misinformed, and a rogues' gallery of those who benefit from sowing social and political chaos. Vaccine makers and governments had to combat a never-ending stream of falsehoods and misrepresentations, some of which came from those governments' own politicians.

The final chapter looks to the future and toward other opportunities for solving big challenges facing humanity. The mRNA vaccine technologies may hold the key to a new era of rapidly developed and effective vaccines for a wide range of diseases. Moreover, the ability to program a patient's cells to make therapeutic substances could address a wider array of health problems such as cancer and organ failure. The combinations of genetic technologies and vaccine mass-production technologies can also be applied beyond medicine and agriculture to industrial processes. Finally, the deeper processes of accumulating scientific understanding, creating portfolios of engineering solutions, and building large-scale ecosystems of manufacturing and supply chains can combine to address other global problems such as climate change.

In summary, this book is a tale of bringing the full might of science, engineering, supply chain processes, and government resources to combat a critical global problem. Each of these four realms of human endeavor faced, and largely overcame, serious obstacles in pursuit of the goal of preventing more death, disease, and economic upheaval from

Covid-19. Overall, the great race to vaccinate humanity holds many lessons about product development, manufacturing, creating new supply chains, distribution, and customer adoption of highly innovative, revolutionary products.

Acknowledgements

Like my last book, *The New (Ab)Normal: Reshaping Business and Supply Chain Strategy Beyond Covid-19* (MIT CTL Media, 2020), this book was written while its content (the vaccination drive) was still unfolding. The compressed time frame for research and writing was made possible owing to the help of an incredibly dedicated and talented team. First and foremost, my thanks to Andrea and Dana Meyer, the team behind Working Knowledge®. They were responsible for researching and organizing the information, writing and editing drafts, and—most importantly—served as a tough sounding board for ideas and approaches to the coverage.

The marketing team at the MIT Center for Transportation & Logistics helped in multiple ways. Dan McCool was a tough editor who made valuable suggestions, as did Ken Cottrill. Dan was also responsible for designing the look of the pages and typesetting the text. Arthur Grau came up with the cover design, and Emily Fagan helped in the process of self-publishing. Elizabeth Hamblin proofread the text and organized the endnotes. My deep thanks to all of them.

The book benefited from interviews with several colleagues, researchers and executives who were very generous with their time:

Marcello Damiani, Chief Digital and Operational Excellence Officer at Moderna

Gil Epstein, Professor of Economics and Dean of Social Science at Bar-Ilan University

Paul Granadillo, Senior Vice President of Supply Chain at Moderna

Howard M. Heller, Senior Advisor for Clinical Partnerships at the Massachusetts Consortium on Pathogen Readiness at Harvard Medical School, and Infectious Disease Consultant at Harvard Medical School and Massachusetts General Hospital

Anette "Peko" Hosoi, the Neil and Jane Pappalardo MIT Professor of Mechanical Engineering and Associate Dean of the MIT School of Engineering

Katalin Karikó, Senior Vice President at BioNTech RNA
Pharmaceuticals and Adjunct Professor at the University of
Pennsylvania

Robert Langer, Biomedical Engineer, Institute Professor at MIT, and
Co-Founder of Moderna

Phillip Sharp, Nobel Prize-Winning Biochemist and Institute
Professor at MIT

Meri Stevens, Worldwide Vice President of Consumer Health Supply
Chain and Deliver at Johnson & Johnson

1. Creating New Vaccines in Record Time

Whenever a new disease lands on humanity's doorstep, epidemiologists wonder whether it will become a footnote in an obscure academic article or a history-making global plague. They immediately ask, *Is it transmissible from human to human? Who is susceptible? How contagious is it? And how bad is the disease in terms of health effects, hospitalizations, and deaths?*

If the disease reveals itself to be contagious (i.e., exhibiting a high reproduction number, R_0, which is the average number of people in a susceptible population who will contract the disease from each person with that disease) and devastating to health (i.e., a high infection fatality rate), then the epidemiologists start to consider their responses, such as:

- Can the disease be contained so that very few people get sick (i.e., ensuring that $R_0 < 1$ so the infection will die out over time)?

- Can doctors easily treat those who are infected to avoid serious illness and death (e.g., the old nemesis of bubonic plague is now easily cured with antibiotics)?

- Can a vaccine immunize the susceptible in order to forestall sickness and, ideally, halt the spread of the disease?

In the case of Covid-19, which proved to be both highly contagious and deadly, few countries had the leadership, the means, and the willing participation of their citizenry to contain the disease. A huge part of the containment challenge was that a large percentage of infected people showed no symptoms, making them difficult to identify and isolate in order to prevent further spread of the disease. Worse, doctors found that the available treatments were not very effective; even patients in the most advanced hospitals still died. That left vaccination as the last best hope for defeating Covid-19 before it killed untold numbers of people.

Prior to Covid, scientists typically needed 10–15 years to develop a new vaccine and establish its safety and efficacy.[1] Worse, vaccine development historically suffered from long odds of success even after those years of dedicated effort—for example, of all vaccines developed from 1998 to 2009, only 6 percent of candidates made it to market.[2] The fastest previous vaccine development project in history had been for mumps. In 1963, scientists at Merck leveraged decades of vaccine technology improvements to create a new vaccine that debuted in 1967.[3] The long odds and previous record of a four-year development process fueled pessimism as scientists embarked on a race for a vaccine against the SARS-CoV-2 virus[*] in early 2020. What the pessimists did not realize, however, was that all the scientific progress of the intervening decades would serve as the foundation for what happened next.

The Long Road to Quick Vaccine Development

The word vaccine derives from *vacca,* the Latin word for cow, and it arose from the invention of the first widely manufactured vaccine. In the late 1700s, Dr. Edward Jenner, an English physician, proved that people who caught cowpox—a mild disease of cows sometimes transmitted to milkmaids—did not catch smallpox, which was a deadly and disfiguring disease that routinely killed hundreds of thousands of people yearly. In a 1798 report, Jenner concluded that "the cowpox protects the human constitution from the infection of smallpox."[4] His approach of exposing a person to a weak, inactivated, or killed variant of a dreaded disease became a key approach to immunization that is used to this day. In fact, several Covid vaccines use this weakened or attenuated pathogen strategy.

---

[*] SARS-CoV-2 is the name of the virus that causes the disease Covid-19.

Discovering the Essence of Vaccination

Dr. Jenner made clever use of the observation that cowpox gave milk-maids immunity to smallpox, but he could not have known exactly why this worked. Decades of slow and steady research would be needed to discover that microorganisms can cause disease, that the human immune system can recognize these foreign invaders, and that foreign substances (foreign proteins) associated with the invaders act as anti-gens,[†] sparking production of antibodies to more quickly fight subsequent infections.[5]

Finally, after more than a century of research into immunology, scientists realized they didn't need to use the whole virus or bacterium to make a vaccine. Like teaching children to find Waldo[‡] by simply looking for his iconic red-and-white striped pullover, vaccines can train the immune system to recognize pathogens simply from some emblematic element of the pathogen's external appearance, such as the proteins that coat a viral particle or bacterial cell.

Thus, a vaccine could be made from just Waldo's red-and-white sweater without including the living and potentially dangerous Waldo. In the case of Covid-19, the spike protein is Waldo's distinctive sweater that alerts the body to the presence of the virus, thus immunizing against the disease. That finding led to a second vaccine technology based on synthesizing and purifying proteins or fragments from the pathogen to be used as a vaccine. Injecting these noninfective substances would then trigger antibody creation that would be ready to attack the target pathogen if (or when) the vaccinated person was exposed to the disease.

Learning the Secrets to Life's Building Blocks

While one set of scientists was discovering the essentials of immu-nology, others were exploring the nature of heredity and genetics to understand what determined the building of proteins in the human

---

[†] *Antigens* are foreign substances that induce immune response in the body, especially the production of antibodies.

[‡] *Where's Waldo?* (*Wally* in Britain) is a series of children's puzzle books con-sisting of illustrations depicting throngs of people at a given location. Readers are challenged to find a character named Waldo hidden in the group. Waldo is identified mainly by his red-and-white-striped sweater, which is disguised in the illustrations.

body and other organisms.⁶ The road to scientists' understanding of the building blocks of life probably started in 1869, when Swiss researcher Friedrich Miescher was first to identify DNA (deoxyribonucleic acid) as a distinct molecule. In 1881, the German Nobel laureate Albrecht Kossel not only gave the molecule its DNA moniker, but he also identified its four nucleic acid building blocks. In 1953, molecular biologists James Watson and Francis Crick discovered the double helix structure of DNA as a long string of nucleotide letters.⁷ In time, scientists learned how strings of the four nucleotides§ of DNA formed a language for storing recipes that translated into sequences of the 21 amino acid building blocks of proteins, enzymes, and other functional elements of life.

Then, in 1961, scientists discovered mRNA (messenger ribonucleic acid), which comprises shorter, single-strand strings of nucleotides that encode the recipe for just one of a cell's proteins or enzymes.⁸ Whereas a cell's very long strings of DNA remain locked inside the nucleus, mRNA leaves the nucleus and enters the bustling bulk of the rest of the cell, where it is quickly translated into the individual strings of amino acids. If DNA is a restaurant chef's master copy of all the recipes kept locked in the office, then mRNA is the slip of paper with a copy of the recipe for the soup du jour that the kitchen's sous-chefs will use for that night's cooking (only much more complicated). When a cell needs to make a protein or enzyme, it uses the master copy of the permanent DNA to transcribe multiple copies of these temporary mRNA strings for the needed product, and then each copy of the mRNA gets translated into multiple copies of the end product.

Early on, many scientists recognized the potential power of both DNA and mRNA for a broad range of healthcare applications. Genetic therapies could entice a patient's own cells to make exactly what they needed. For one-time or short-term treatments, mRNA is ideal. A dose of mRNA is like injecting a purchase order (PO); the cell will make some amount of the ordered material as specified in the PO and then naturally stop. Unlike DNA, mRNA normally can't become part of the permanent genetic code of the cell or organism. The temporary nature of mRNA gives it the perfect properties for one-time or short-term

---

§ *Nucleotides* are types of *nucleosides*, which consist of one of the four genetic letters with a sugar group (ribose in RNA or deoxyribose in DNA). Nucleotides are nucleosides with an attached phosphate group that is needed to form the strand backbones of RNA and DNA. Nucleotides form the basic structural unit—the beads—of long-chain nucleic acids such as DNA.

treatments needed for vaccinations, treating a disease, killing a tumor, or repairing tissue damage.

Dr. Phillip Sharp, a Nobel Prize-winning biochemist and professor at MIT, highlighted the advantages of mRNA vaccines. "Messenger RNA vaccines have several benefits compared to other types of vaccines, including the use of noninfectious elements and shorter manufacturing times," he said. "The process can scale up, making vaccine development faster than traditional methods. RNA vaccines can also be moved rapidly into clinical trials, which is critical for the next pandemic."[9] Moreover, the development of treatments based on mRNA and DNA have the potential to end the frustrating trial-and-error process of one-bug-one-drug that has dominated healthcare.

A Fragile Panacea

In theory, the new tools for creating genetic strings meant scientists could program a strand of mRNA to make any sort of vaccine antigen or therapeutic protein or enzyme. For example, the mRNA that encodes the Covid spike protein could make an excellent candidate for a Covid vaccine. In practice, nothing in biology or healthcare is ever so simple. The human body has an elaborate system of layers of border control and monitoring processes in its organs and cells to keep suspicious, foreign, and dangerous substances out. While scientists were hopeful about injecting mRNA into human cells, the cells themselves responded to the first experiments with innate immune defenses against foreign RNA.

The high immunogenicity[†] of lab-made mRNA, along with the widespread presence of RNA-destroying enzymes (nucleases), made it hard to get mRNA into cells, and the injected mRNA didn't last long enough to make much protein. For researchers such as Dr. Katalin Karikó, working at the University of Pennsylvania at the time, mRNA's reputation for fragility was a severe hurdle for obtaining funding for her research; few thought that mRNA was a feasible means for medical treatments. In 1997, Karikó teamed up with Drew Weissman, a professor of immunology also at the University of Pennsylvania, to delve into why scientists' mRNA was attacked while the cell's own mRNA was not.

Although Biology 101 drills into every student that there are only four letters of the genetic code in every living organism,[10] the

---

† *Immunogenicity* is the tendency of a foreign substance to provoke an immune response in the body.

biochemistry of these four nucleosides and nucleotides is a bit more complicated than that. Many organisms, including humans, also have chemically modified nucleosides that use subtle molecular variants of the four letters. The modified nucleosides could be considered akin to letters of a word that are italicized or boldfaced—they mean the same thing but look a little different.*

In 2005, Karikó and Weissman discovered that synthesizing mRNA using certain naturally occurring modified nucleosides kept the cell from reacting to the injected mRNAs. Modified nucleosides, Karikó explained, "not only made the RNA non-immunogenic, but we get a lot of protein: 10 times more protein than with conventional RNA."[11] That breakthrough suddenly made mRNA a lot more feasible. Yet, even with this breakthrough, many were skeptical. "We went to biotech companies and pharmaceutical companies to try and get funding, and they weren't interested," Weissman said. "They said RNA was too fragile and they didn't want to work with it."[12]

Although mainstream government research funding agencies and larger pharmaceutical companies still saw mRNA as having too little chance of being feasible, some startups were taking an interest in mRNA and the newfound successes in using it. These included CureVac, BioNTech, and Moderna—the last two founded explicitly to develop mRNA-based treatments based on the discoveries of Karikó and Weissman. In 2013, Karikó moved to join BioNTech in Germany. "I told my husband when I decided to go to Germany, 'I just want to live long enough that I can help the RNA go to the patient,'" she said. "'I want to see... at least one person would be helped with this treatment.'"[13]

Producing Protective Packaging

Proponents of mRNA technology still faced other problems, ones commonly faced by material handlers and distributors in supply chains: "how you deliver [it], because RNA had to be put in something so it is protected, so it won't degrade," Karikó said.[14] An mRNA molecule is a sequence of nucleotide beads on a single, fragile string of molecular bonds. If even one of those bonds breaks, the string doesn't work. That

---

* These modified nucleoside letters have no effect on the basic genetic meaning of the string—the modified mRNA still translates to the same protein product. But on a chemical, structural, and functional level, the modifications can profoundly change the properties of the string.

problem is especially severe in the case of the Covid vaccine, which needs a string of about 4,000 beads. "It's tougher," said Dr. Robert Langer, biomedical engineer, MIT professor, and co-founder of Moderna. "It's a much bigger molecule. It's much more unstable."[15]

Meanwhile, an entirely separate stream of research was exploring how to encapsulate substances inside extremely tiny lipid nanoparticles (LNP)—small bubbles of fats that emulate cellular membranes— as a means of delivering drugs. Initially, using these LNPs for mRNA seemed infeasible, too, because the positively charged lipids that could package mRNA molecules were incompatible with cellular membranes. But decades of research into lipid particles finally found an appropriate class of cleverly designed, pH-sensitive, ionizable lipid molecules.[16] Under acidic conditions, these ionizable lipids become positively charged and can readily encapsulate mRNA during the manufacturing process. Under neutral conditions in the final vaccine and human body, these lipids lose their positive charge and are safe. When a person's cell absorbs an LNP, the cell encapsulates the LNP in a small bubble that becomes naturally acidic, causes the LNP to change conformation,[†] and releases the mRNA. The vaccine makers adopted this technique to package the mRNA and designed it for the needs of getting the vaccines to their intended destinations. "They encapsulate it in a nanoparticle, and that nanoparticle has to have specific properties that, once it's inside a cell, the particle releases the RNA," Sharp said.[17]

As with many supply-chain packaging systems, multiple materials are needed to ensure safe delivery. To further aid the delivery of mRNA, designers added a molecular coating of PEG (polyethylene glycol), which improved the solubility of the oily nanoparticle in water, reduced the body's direct immune system response to the particle, and reduced the chance that the kidneys would remove the particle from the bloodstream and discard it. By analogy, if the lipids are like the bubble wrap that helps protect the mRNA from damage, the PEG is like an outside box that makes the bubble-wrapped mRNA more compatible with the conveyor belts and chutes of the body: the bloodstream.

The result was a drug-delivery technology based on LNP. "It is a tremendous vindication for everyone working in controlled drug delivery," said Langer.[18] Just as specialized packaging companies make and supply bubble wrap and boxes to shippers, specialized ingredient companies

---

[†] In biochemistry, a *conformational change* is a change in the shape of a macromolecule, often induced by environmental factors, such as changes in acidity.

supply ionizable lipids and related technology to vaccine makers.[19] Vaccine candidates from Pfizer–BioNTech, Moderna, and CureVac used modified nucleoside mRNA encapsulated by LNP.

LNPs aren't the only packaging solution to getting genetic material into cells. Nature has had hundreds of millions of years of experience in how to package DNA and RNA into compact, easily shipped capsules— that's exactly what a virus particle is—and some vaccine makers (e.g., J&J; Oxford University–AstraZeneca; Russia's Gamaleya, the maker of Sputnik V; and India's Serum Institute, maker of Covishield) sought ways to use those preexisting shipping materials. Known as a *viral vector* vaccine design, it uses a harmless virus (one that cannot replicate in people) to carry genetic material for the vaccine (e.g., the Covid spike protein gene) into the person's cells.

In this approach, vaccine developers design a string of DNA encoding the target antigen of the disease (e.g., the SARS-CoV-2 spike), splice it into the DNA of a weak or nonreplicating virus, and then mass-produce the modified viral vector to make the vaccine. When injected, the particles of the viral vector infect some of the person's cells, causing those cells to make the SARS-CoV-2 spike protein and prompt the immune system to learn to fight Covid.[20] Thus, the approach updates the original, familiar "weak pathogen" strategy of vaccination with the latest programmable antigen approach of genetic vaccines.

From Discoveries to Tools to Platforms

These stories of biotech development have three broader lessons. The first is that all this progress in biotechnology has been like building a 1,000-piece jigsaw puzzle of the interlocking interactions of life. Each researcher found a piece or two of this puzzle that built upon all the other previously discovered pieces and supported the next generation of discoveries. Some pieces were so important that their discoverers won Nobel Prizes. "This is a field which benefited from hundreds of inventions," said Uğur Şahin, founder and chief executive of BioNTech.[21]

Second, as scientists explored the nature of DNA and RNA, they built new categories of tools to accelerate their work. One category of tools was for sequencing DNA to determine the genetic code of an organism, which enabled scientists to understand the genes of organisms and how different organisms were related to each other. Another category included tools for synthesizing new genetic strings, which enabled scientists to test how different genetic strings created different products and affected the organism. This culminated in programmable laboratory machines that could make DNA or RNA strings with virtually any genetic sequence found or invented by scientists. An essential part of all of this work was the polymerase chain reaction (PCR), which is a way of replicating a tiny sample of genetic material into very large numbers of copies that can be used in experiments, tested for certain properties (e.g., "Does this tiny sample contain the Covid virus?"), or packaged as a product. The final key technology was reverse transcriptase, an enzyme that translates a string of messenger RNA (such as from a virus) back into a string of DNA that can be readily copied for testing or mass production. All these tools involve global supply chains for laboratory equipment, supplies, chemical reagents, and services (more on this in Chapter 2, pp. 23–29).

The third story is the concomitant development of digital and biological technology platforms to support new product development using all this accumulated knowledge. For example, Moderna "decided from the beginning to build from the ground up a digital biotech," said Marcello Damiani, Moderna's chief digital and operational excellence officer. "We made sure that the company is data-centric and that we can get insights from this database to help us improve quality and efficiency and accelerate our learning." With a digital system, Moderna can boost researcher productivity from testing 40 mRNAs per month to 1,000. "This gives you an idea how the digitization and the use of this technology can enhance drastically the efficiency of the scientists," Damiani said.[22]

These Genes Are in Fashion

When Chinese scientists first isolated the virus that caused Covid-19 and identified it as a coronavirus on January 7, 2020, it provoked fear but also created an opportunity. The fear arose from the experiences the world had with two previous novel coronaviruses. In November 2002, severe acute respiratory syndrome (SARS) emerged in China as an outbreak of a severe and contagious respiratory disease that infected more than 8,000 people and killed 10 percent of them before it was contained in mid-2003.[23] Next, in 2012, Middle East respiratory syndrome (MERS) appeared,[24] and, although less contagious, it was much more lethal (35 percent of patients with MERS died).[25] Both of these coronaviruses were quickly contained by quarantines, contact tracing, and the relatively poor effectiveness of SARS and MERS in spreading from human to human.

Dusting Off Old Data

With the SARS-CoV-2 virus that causes Covid-19, scientists had an opportunity to use the knowledge gleaned from analyzing SARS and MERS. This included initial efforts at developing a vaccine for SARS, which stalled due to lack of funding when the SARS outbreak ended.[26] SARS and MERS also stem from coronaviruses, as does SARS-CoV-2. Note that the coronavirus gets its name for the *corona,* or crown of spikes that covers the outside of the virus. The corona effectively makes the pathogen into a Velcro ball that can stick to certain receptors on cells in people's lungs, blood vessels, intestines, and elsewhere. Those spikes are the dominant outermost feature of the virus that antibodies and immune cells are most likely to be able to find, recognize, and latch onto. The work on the earlier coronaviruses made it clear that this spike protein was the obvious antigen for Covid vaccine development—the equivalent of Waldo's sweater.

Chinese researchers published the first data on the genetic sequence of SARS-CoV-2[27] on January 12, 2020.[28] As Covid hurtled toward pandemic status,‡ thousands of researchers in both private and public organizations headed to their labs. Data, fresh insights, and research papers streamed across the internet. The Allen Institute used Semantic Scholar, its AI-based search engine,[29] to build the Covid-19 Open Research Dataset,[30] which organized more than 130,000 scientific articles related to the virus by September 2020.[31] Dr. Rena Conti, associate research director of biopharma and public policy at Boston University's Questrom School of Business, described the effort: "We've watched physicians, scientists, and technologists completely step up to the challenge here and promote information sharing, promote good practices, promote messaging around what's effective, what's safe, and what's possible."[32]

The challenge was to quickly use the tools of science and biopharmaceutical manufacturing that had been accumulating for more than 150 years. Moderna exemplifies the new breed of vaccine creators using genetic technologies to accelerate development. Within two days of

---

‡ The World Health Organization (WHO) declared Covid-19 a global pandemic on March 11, 2020.

getting the genetic sequence of the virus, Moderna had designed its mRNA vaccine.[33] Within about a month, the company had manufactured, tested, and shipped samples of its vaccine to the National Institutes of Health (NIH).[34] More broadly, within the first four months of the arrival of Covid-19, research groups around the world were developing 115 different candidate vaccines.

Vaccines on Trial for Your Life

All the accumulated knowledge, advanced technologies, and a deep understanding of virology and immunology helped create those 115 candidate vaccines. But that knowledge offered no guarantee that any candidate would work inside a complex, messy human body. Before being injected into millions, perhaps billions of arms, a vaccine—like any pharmaceutical product—must be carefully tested in clinical trials. These trials assess whether the candidate vaccine helps people avoid the disease or reduces its impact and whether it is safe in not having serious side effects.

By modern standards of sensibility, law, and ethics, history's first vaccine trial was appalling. In 1796, the aforementioned Dr. Edward Jenner transferred pus from a milkmaid's cowpox pustule into a cut on an eight-year-old boy's arm and then subsequently exposed the boy to smallpox.[35] This boy and Dr. Jenner's other test subjects were very lucky; they survived and proved to be immune to repeated exposures to smallpox. As medical science and ethics improved, scientific and regulatory developments sought to eliminate much of the risks when it came to exposing human subjects to new medical treatments such as vaccines.[36]

Arsenal of Assessment Methods

The key to safer, more ethical approaches to creating vaccines and other drugs came from developing a portfolio and sequence of unbiased testing procedures designed to check for safety and efficacy of any new proposed treatment. Testing a new drug or vaccine starts with *in vitro* studies (Latin for "within glass") that test new therapeutic agents in test tubes and petri dishes on cells and tissue cultures for disease-related efficacy and toxicity. Scientists have developed a large number of very carefully managed cell and tissue culture lines to ensure repeatable results that are comparable across labs.

Next come *in vivo* ("within living creatures") tests, typically on animals such as mice and monkeys. Scientists try to find animals that are susceptible to the target disease and exhibit similar symptoms (e.g., SARS-CoV-2 sickens hamsters),[37] which can be used to test vaccines and treatments. These laboratory testing efforts rely on their own well-developed supply chains of providers of cells, tissues, lab animals, and the myriad of specialty chemicals used to detect, assess, or visualize relevant biological effects.

The advent and increasing power of computers and the accompanying software has also led to *in silico* ("in silicon") computer testing. Scientists can create and use mathematical models or simulations of drug molecules, metabolic pathways, biological proteins, and genetic sequences to predict drug performance, disease progression, potential side effects, and other interactions. For example, *in silico* models for molecular docking, molecular dynamics simulations, network-based identification, and comparative modeling can help repurpose already-approved drugs that, theoretically, might be effective against a new disease.[38] Because the safety of these drugs has already been proven, only their efficacy need be established. All these possibilities mean that scientists have an ever-growing pool of knowledge and an arsenal of testing methods to predict, test, and confirm the safety and efficacy of new vaccines and drugs.

The First Valiant Volunteers

If a candidate vaccine or other drug survives preclinical trials in computers, test tubes, and lab animals, it moves into a tightly regulated, multi-phase, clinical testing process with human subjects. In the US, that process begins with an Investigational New Drug (IND) application,[39] in which the maker tells the Food and Drug Administration (FDA) exactly what it is doing, provides preclinical evidence that the product will work, and states the specific measures of performance they seek to show. (In the case of Covid vaccines, this includes showing the reduction in infections, hospitalizations, and deaths.) The FDA must approve the application before human trials can begin.

Requiring this preapproval ensures that proponents of a new treatment don't run multiple trials in secret and then only submit the results that have a favorable outcome. After-the-fact cherry-picking of data or results would let ineffective products appear to be effective. Like a billiard player being required to call out which ball they plan to sink into which pocket, the IND helps prove that any results in a clinical trial are real and not just luck.

After approval of the IND application, human clinical trials for new vaccines (and other drugs) have at least three phases.[40] Phase 1 tests different doses of the candidate vaccine on several dozen healthy subjects, watching for side effects and testing for the vaccine's effect on the immune system. Phase 2 more carefully tests several hundred subjects for dosing and efficacy using a controlled, randomized, and double-blind protocol in which neither the doctors nor the patients know who is getting the vaccine or a placebo. Phase 3 tests for efficacy and safety in an even larger group, running into the tens of thousands of people, as in the case of the Covid vaccine trials.[41]

By April 2020, at least 18 vaccines were in preclinical testing and five were already in Phase 1 trials on humans.[42] For example, Moderna started its Phase 1 trials in mid-March 2020 with plans to enroll up to 155 healthy adults ages 18 or older, testing five dosage levels (10 to 250 micrograms of mRNA), injecting only few subjects at a time, and monitoring the results for any signs of adverse effects.[43] At the end of May 2020, the company began a 600-person Phase 2 trial (healthy subjects, including a cohort of older subjects) using the two most promising dosage levels.[44]

As the months ticked by and Covid cases surged, dozens of hopeful vaccine developers saw encouraging results from their Phase 1 and Phase 2 trials. Their candidate vaccines seemed safe, and subsequent blood tests showed that their vaccines had triggered the subject's immune system to make antibodies. However, and it was a big however, a vaccine's ability to create antibodies was no guarantee that the subject's immune system would successfully fight off an infection in the real world. The true test would come in Phase 3, which provides the final set of data needed for regulators to decide whether to approve the new product or not.

The Ethics of Exposure

Phase 3 trials test a much larger group of people to determine if the treatment or vaccine really works as the maker expects and as the regulators demand. In the case of drug trials for treatments of people who have a disease, the trials can readily enroll a few hundred test subjects with that disease and immediately see if the treatment helps. In contrast, trials for vaccines must enroll healthy people, randomly vaccinate some of them (with the rest receiving a placebo), and then somehow expose the subjects to the pathogen. Clinicians have two choices for this exposure step: either a challenge trial (by intentionally trying to infect people) or natural infection (by unintentional exposure as the subjects go about their lives).

Unlike Dr. Jenner with his challenge trials for a cowpox vaccine in the 18th century, modern-day clinicians are loath to intentionally attempt to infect people with a disease if the infectious disease has a chance of severe consequences and no known adequate treatments, as in the case of Covid-19. However, without the use of challenge trials, the trial's managers must wait for the test subjects to catch Covid naturally. That can take time depending on the prevailing levels of Covid in the community and the test subject's physical-distancing habits. Such trials cannot deliver results until enough subjects (of unknown vaccination status) catch the disease. Only then can the researchers unblind those subjects' data to see if those who caught it were vaccinated or not. That implies the vaccine maker must either enroll very large numbers of subjects or wait a very long time for the results. For example, Moderna's Phase 3 clinical trials enrolled more than 30,000 test subjects and planned to wait until at least 160 people caught the disease.

Some ethicists argued for (voluntary) challenge trials, when faced with a pandemic like Covid, on the grounds that many more people would catch Covid and die while waiting for the slower non-challenge trials to progress than they would were challenge trials used.[45] In theory, challenge trials could provide solid data in as little as two months, whereas the standard ones took more than four. (On the flip side, a challenge trial may suffer from systemic bias to the extent that the method of exposing test subjects to Covid is either stronger or weaker at triggering infection compared to how people are normally exposed to the virus in everyday life.)

In August 2021, the UK did use Covid human challenge trials in order to understand how the virus attacks the body from the moment of exposure. Thirty volunteers aged 18–30 were checked, hospitalized, and infected, allowing scientists to take multiple measurements daily of viral load, blood composition, urine and stool content, and antibody levels. "Because we can take so many different samples, we can get extraordinary insight into how the virus causes disease," said Dr. Peter Openshaw, a professor of experimental medicine at Imperial College London and a co-investigator on the study.[46]

In the absence of challenge trials for the vaccines, the higher numbers of test subjects required per trial substantially increase the costs of the trials. The reason is that each test subject might undergo dozens of clinical visits and medical tests to assess their health, receive the vaccine, and be periodically checked for both side effects and the vaccine's efficacy. Moreover, clinical trials also typically compensate test subjects for their time and effort—a few hundred to a few thousand dollars per subject in the US.[47]

The Real Test in the Real World

When the leading vaccine candidates started Phase 3 trials in the late summer of 2020, the timeline for obtaining sufficient data in the trials depended on how fast Covid was spreading wherever the test subjects lived. Ironically, the inability of governments to control the spread of Covid (and of citizens to comply with physical distancing practices) in the fall of 2020 aided these vaccine trials. The poorer a government's control of Covid case rates, the higher the chances that the test subjects would be exposed to Covid and form a noticeable pattern: more cases among the placebo-injected control group and fewer cases in the vaccine-injected treatment group. "The number of cases did help expedite the performance of clinical trials, both here and abroad," said Stephen Hahn, former commissioner of the FDA. "That did help us get to this point. But I think none of us would have wanted it to be that way."[48]

Moderna's Phase 3 trial for its mRNA vaccine began at the end of July 2020, with 30,000 volunteers spread out at 89 sites around the US[49] The trial sought healthy adults (including those with preexisting conditions as long as they were medically stable) with an emphasis on people with a high risk of Covid infection due to their location or circumstances.[50] Pfizer launched its own Phase 3 trials around the same time, with some 44,000 volunteers in 39 US states as well as Brazil, Argentina, and Germany. In general, regulators encouraged people in vulnerable medical and racial groups to volunteer for Covid vaccine trials to ensure that any approved vaccine would be effective for them.[51] By June 2021, nearly three dozen vaccines had done well enough in early trials that the makers were pursuing Phase 3 trials.

Some potential subjects were excluded from the trials. In general, ethical concerns preclude initial testing of new medical products on children and on women who are pregnant or breastfeeding (unless these groups are especially prone to the disease). In the case of Covid, that risk-averse ethical principle did have the unintended consequence that no one truly knew if any Covid vaccine was safe or effective in these two groups.[52]

Separate, follow-on trials with these groups typically occur after successful trials with other adults. For example, Moderna started a follow-on Phase 3 trial with almost 4,000 adolescent children (aged 12–17) in December 2020.[53] Meanwhile, Pfizer started a follow-on combined Phase 2 and 3 trial starting with 350 pregnant women and expanding to 4,000 in February 2021.[54] As of the writing of this book, younger children (under 12) remain wholly unvaccinated because clinical trials for younger ages have not been completed. Fortunately, on September 20, 2021, Pfizer announced that its tests have shown that its Covid-19 vaccine works well for children 5–11 years old. This vaccine may be approved later in 2021.

And the Winner Is…

To attain approval, authorities specified the minimum required effectiveness of a Covid vaccine to be at least 50 percent (cutting the chance of infection or serious illness by at least half).[55] "I know that 50 percent does sound low, but that is still some protection, and some protection is better than no protection," said Dr. Jeff Kwong of the Centre for Vaccine Preventable Diseases at the University of Toronto.[56] Although 50 percent effectiveness wouldn't immediately eliminate Covid, it would likely cut the death rate significantly in vulnerable populations, reduce the burden on healthcare systems, permit some relaxation of economy-crushing physical distancing policies, and help reach herd immunity levels (in which Covid becomes an occasional local nuisance rather than an ongoing global menace).

The Results Roll In

As mentioned above, for Phase 3 trials that rely on volunteers' natural risk of exposure to Covid, researchers must wait until a sufficient number of trial subjects have caught Covid before they unblind the data. If the vaccine is 100 percent effective, then none of those who were infected with Covid will be from the vaccinated group. If the vaccine is 50 percent effective, then the fraction of the vaccinated group who caught Covid will be about half that of the placebo group. Related trial data on hospitalizations and deaths will show if the vaccine helps avoid serious disease even if it doesn't prevent every infection. Data on side effects among the vaccinated and unvaccinated will help estimate the safety profile of the vaccine. Regulators then use that data to determine whether to authorize the use of the vaccine or not.

When Pfizer analyzed the first 170 Covid cases in its 43,000-person Phase 3 trial, the vaccinated group had only eight cases, while the placebo group had 162.[57] The Pfizer–BioNTech mRNA vaccine was therefore estimated to be 95 percent effective. "We were overjoyed," said Dr. Ann Falsey, a professor of medicine at the University of Rochester, who ran one of the Pfizer vaccine trial sites. "It seemed too good to be true. No respiratory vaccine has ever had that kind of efficacy."[58] Similarly, when Moderna analyzed the first 196 Covid cases among its trial's participants, the results were similarly promising: Only 11 of the cases were among the vaccine recipients, while 185 were among the placebo group (94 percent effectiveness).[59]

Regulators Rule

In mid-December 2020, the FDA issued Emergency Use Authorizations (EUA) for Pfizer's and Moderna's respective mRNA vaccines within a week of each other. Numerous vaccines obtained some sort of regulatory approval in the latter half of 2020 and first half of 2021. Russia approved its Sputnik V vaccine in August 2020 based solely on favorable Phase 1 and 2 data before even conducting Phase 3 trials to determine if the vaccine actually prevented the disease.[60] By October 2020, five vaccines were already approved for early use in some countries, 11 were in large-scale Phase 3 trials, 42 were in small-scale human trials, and 91 vaccines were in preclinical stages of development.[61] Some vaccines fell by the wayside. For example, Merck abandoned development of two candidates when early tests showed lackluster immune response.[62]

Other vaccines had mixed effectiveness among different age groups. Oxford–AstraZeneca's vaccine seemed more effective in those over 65 (85 percent efficacy against symptomatic Covid) than overall (76 percent efficacy).[63] On the other hand, CanSino, a Chinese viral vector vaccine, seemed to have less effectiveness in older subjects compared to younger ones. (This was likely due to those older subjects having had some exposure to the vaccine's adenovirus[§] vector sometime during their long lives.[64] Their immune systems likely fought off the virus vector before it had time to make enough of the Covid antigen). But even if a vaccine lacked effectiveness in all groups, it could still be given to those for whom it did work well. Given the bottlenecks in vaccine production, every dose could make a difference in helping gain global herd immunity and forcing the retreat of Covid.

By early August 2021, 21 vaccines were approved for use in at least one country—several of which were approved in dozens of countries—and another 41 had done well enough in early tests to justify Phase 3 trials.[65] At that time, the Oxford–AstraZeneca was the most widely approved vaccine—usable in 121 countries.[66]

Best Supporting Actor in a Major Pandemic: Science

Less than a year after identifying the SARS-CoV-2 virus as the cause of Covid-19, both the United Sates and the European Union had approved both the Pfizer–BioNTech and Moderna mRNA vaccines: the first ones usable in those two jurisdictions.[67] Dr. Eric Daar, chief of HIV Medicine at Harbor–UCLA Medical Center, explained the fast development: "The reason we are here, less than a year into the pandemic and we have two effective vaccines, is because of what came before this—efforts to develop vaccines for HIV or MERS or SARS or Ebola. People developed these novel strategies while pursuing vaccines for these other diseases, so they were basically on the lab bench waiting for the next pandemic to come along."[68]

Dr. Steven Joffe, professor of medical ethics and health policy at the Perelman School of Medicine at the University of Pennsylvania, added, "We are lucky in the sense that the science was there."[69] These 21 approved vaccines included three mRNA vaccines, six viral vector vaccines, eight inactivated virus vaccines, and four protein subunit vaccines.[¶] All these efforts cost billions of dollars.

[§] An *adenovirus* is a member of a family of viruses often associated with the common cold. Some vaccine makers used genetically engineered versions of these types of viruses and modified them to not replicate and sicken people while carrying the genetic instructions for the antigen.

[¶] Subunit vaccines, unlike inactivated whole-cell vaccines, contain only the antigen part of the pathogen.

Global Funding

A loose worldwide collection of governments, international health agencies, rich benefactors, and deep-pocketed pharmaceutical companies supported the race toward worldwide vaccination. They anted up billions of dollars to spin the giant roulette wheel of high-stakes vaccine development with hopes of getting first dibs on any resulting vaccines. Given the stakes in lives and livelihoods, many players bet big.

Funding the Doubly Risky Double Parallel Strategy

Many of these players pursued "double parallel" development strategies to increase their chances of success and decrease their time to production. First, they funded multiple, parallel, competing vaccine candidates to increase the likelihood that at least some of them would work. While that strategy increased the chance that one or more of the efforts would yield a safe and effective vaccine, it risked overfunding the effort and potentially creating more competition for scarce supplies in vaccine supply chains (described in detail in Chapter 2). Second, each effort included funding, in parallel, both the development stage and the production stage rather than the traditional sequential approach, in order to reduce the time from lab to jab. This risked wasting money on capacity that could not be used if that vaccine candidate failed.

Double parallel funding is like playing multiple hands of poker at the same time and going "all in" before getting to look at the cards; some waste is inevitable. However, compared to the impact of millions of lost lives and the trillions of dollars lost owing to curtailed economic activities, it was a small price to pay in order to maximize the chance of delivering billions of doses of safe and effective vaccines as soon as possible. Furthermore, the risks were limited because failed vaccine efforts would have had a minimal impact on vaccine supply chains since they would not continue to procurement and manufacturing at scale. In addition, any manufacturing capacity developed for a failed vaccine could then be adapted to a successful one or used for some other therapeutic product.

An example of this double parallel strategy was the United States's Operation Warp Speed, which was announced in mid-May 2020. It was designed as a public-private partnership with almost $10 billion in initial funding to accelerate the development, production, and distribution of vaccines, treatments, and diagnostics for Covid-19.[70] Operation

Warp Speed negotiated deals for vaccine development and production with manufacturers that included Moderna ($2.5 billion), a Sanofi-GlaxoSmithKline partnership ($2.1 billion), Novavax ($1.6 billion), Johnson & Johnson ($1.5 billion), and AstraZeneca ($1.2 billion). Later, a $2 billion vaccine supply contract with Pfizer covered production only.[71] "Each is coming with a lot of prior experience," said Rick Bright, former director of the Biomedical Advanced Research and Development Authority (BARDA), a federal agency that funds disease-fighting technology. "They all know how to scale up."[72]

In total, Operation Warp Speed secured up to 800 million doses of six potential vaccines, with options to buy up to another 1.6 billion doses. Although these totals vastly exceeded the roughly 660 million doses needed to vaccinate 100 percent of the US population (assuming two doses per person), procuring an oversupply would ensure that even if some vaccines failed to get approval or suffered delays in production, the US would still be able to vaccinate the entire country as fast as possible.

Funding Vaccines for the Underfunded

While the US crafted America-centric investments and deals, a large coalition of countries, international agencies, and philanthropic groups formed COVAX: a combination of a risk pool for vaccine development, a buyer's club for vaccine procurement, and a philanthropic vaccine distribution effort. This worldwide coalition was co-led by GAVI (formerly known as the Global Alliance for Vaccines and Immunization), CEPI (Coalition for Epidemic Preparedness Innovations), and the World Health Organization (WHO).[73] The COVAX model replaced the risky and disjointed process of each country trying to place bets among the contending early-stage vaccine candidates and then fighting to ink deals for favorable vaccine access. COVAX also made equal access for 92 middle- and lower-income countries a cornerstone of its model, on grounds of fairness and on the pragmatic basis that as long as Covid raged in some countries, it would remain a threat to global health, global security, and the global economy.[74] As of early 2021, the coalition had $7.5 billion in funding.[75]

From its beginnings in 2000, the Bill and Melinda Gates Foundation has always focused on vaccination as a key tool to fight infectious diseases in the developing world. In 2010, it pledged $10 billion over 10 years to help research, develop, and deliver vaccines for the world's poorest countries.[76] Long before Covid, the foundation made investments in or grants for mRNA vaccines—totaling more than $150 million—to three of the leading mRNA biotech companies: BioNTech, Moderna, and CureVac. (Those investments both funded development and contained contract clauses aimed at improving global access to any vaccine developed with the Gates Foundation's money.) By the end of 2020, the Gates Foundation had pledged $1.75 billion to the fight against Covid.[77] Without its efforts, the Covid-19 crisis would have almost certainly been worse. "We're a stopgap, we're an accelerator, we're a catalyst," said Gates Foundation CEO Mark Suzman.[78]

Big Pharma Taps Its Own Deep Pockets

Some pharmaceutical companies, such as Pfizer, eschewed government investment for development. Pfizer CEO Albert Bourla explained, "When you get money from someone, that always comes with strings. They want to see how we are going to progress, what type of moves you are going to do. They want reports. I didn't want to have any of that."[79]

Instead, the company invested $2 billion of its own cash to move fast in developing a vaccine using BioNTech's then unproven mRNA vaccine technology.[80] "Usually, we do things sequentially because otherwise it can become very risky, very expensive," Bourla said. This time, however, the CEO told workers, "Don't think like that. Think in parallel, not sequentially.... This is not business as usual. Open the checkbook. Just do it."[81]

The company's big bet and parallel efforts paid off. The Pfizer-BioNTech vaccine became the first Covid vaccine approved for emergency use in the US[82] and the EU.[83] By the end of July 2021, the vaccine had been approved in 97 countries.[84] At that time, Pfizer was forecasting that the vaccine would bring in $33.5 billion in revenue in 2021.[85]

From Trials to Vials

"Creating a vaccine is only half the battle," Bill Gates wrote in a *Washington Post* opinion piece on March 31, 2020.[86] "To protect Americans and people around the world, we'll need to manufacture billions of doses.... We can start now by building the facilities where these vaccines will be made." He should also have mentioned the need for distribution and vaccination sites once vaccines are manufactured. To that end, in August 2020, Operation Warp Speed executed a previously signed contract option to make McKesson the central distributor for Covid vaccines and supplies in the US[87] In early 2021, the US brokered a deal to pay Merck $268.8 million to convert two manufacturing sites to help a competitor, Johnson & Johnson, make more vaccine doses.[88] While researchers and doctors worked feverishly in their labs and clinics, factory managers and workers raced to create the capacity to produce billions of doses of the leading vaccines in record time.

2. From Creation to Quantity

After scientists formulate, clinically test, and successfully gain approval for any new pharmaceutical product, the next step of the typical product development path has teams of engineers and supply chain experts take over to mass-produce and distribute the approved product. Meri Stevens, a global supply chain leader in Johnson & Johnson's consumer health business, said about J&J's vaccine, "Initially it was very much about the science and discovery, but very quickly we were having to create whole cold chains that didn't exist before."[1]

The Science (and Regulation) of Scale

In addition to the heroic feats of medical science in inventing a safe and effective vaccine for Covid-19, vaccine makers had to create supply chains and manufacturing capacity to produce hundreds of millions, and then billions, of doses. And as with the scientific development of the vaccines, the engineering efforts to mass-produce the vaccine were shaped by history. In particular, vaccine manufacturing built on decades of cumulative progress in chemical manufacturing, pharmaceutical supply chain development, and regulation of the associated processes.

From Cooking to Chemistry

The first pharmaceuticals were natural products like herbs, plant extracts, animal parts, oils, and mineral supplements that people believed could relieve suffering, treat illness, and extend life. Early pharmaceutical makers borrowed the techniques of cooking, such as measuring, mixing, soaking, heating, cooling, filtering, and dividing to convert supplies of raw materials into doses of desirable nostrums. The brewing of beer (using microorganisms to convert raw ingredients into desirable products) and distillation of spirits and perfumes (separating valued materials from other substances) added more methods to the early pharmaceutical maker's portfolio of tools.

In the modern era, while biologists pursued an understanding of living organisms to enable the creation of new drugs and vaccines, chemists pursued an understanding of how to identify and synthesize substances, and chemical engineers invented ways to make specific substances in greater quantities, of better purities, and at lower costs. Each new invention increased the range of molecules that people could make and increased the options for reaching affordable large-scale production.

The accumulated body of knowledge in chemistry, biochemistry, and chemical engineering provides pharmaceutical companies and their suppliers with a set of scalable processes and technologies for synthesizing, purifying, and assembling mixtures of molecules. Scaling up from making just enough doses for a 30,000-subject clinical trial to making enough doses to vaccinate the world—billions of people— means transitioning from test tubes to tanks, from pipettes to pumps, and from manual processes to highly automated ones.

Pharmaceutical factories look very different from those of most consumer products. For consumer products such as computers and cars, factory workers and robots physically assemble the products on meticulously organized assembly lines in humming factories. In contrast, pharmaceutical workers use the invisible phenomena of biochemistry to assemble the molecular chains of nucleotides of mRNA vaccines in a soup of raw materials and enzymes in modestly sized tanks sitting quietly in wheeled racks in sterile white rooms. Consumer product factories carefully choreograph the actions of each worker or machine to assemble each part, whereas chemical and pharmaceutical product factories rely on the massive parallelism of near-random molecular interactions inside the tank. Specifically, trillions upon trillions of molecules accidentally bumping into each other thousands of times per second leads to the cleaving and merging of the molecules to synthesize the desired substance. In designing a pharmaceutical production process, the chemist's art is in knowing which raw material molecules and conditions will be most effective at yielding the desired intermediate or end product with the lowest cost and smallest quantities of byproducts.

Supplying Vaccines to the World from a World of Suppliers

The shift from invention to production also entails a shift in required suppliers. Making billions of units of a product involves tapping into global industries of suppliers who can produce large volumes of ingredients and intermediate products in their factories. If invention is about accessing the right ideas, production is about accessing the right capacity.

Pharmaceutical supply chains started becoming global almost from the beginning; the natural products of the first pharmaceuticals often only grew in or came from distant lands. For example, biblical myrrh (used as an anti-inflammatory and antibacterial agent as well as incense) came from sap harvested from a particular thorny shrub native to the Horn of Africa and was exported throughout the ancient world.[2] This tendency toward global sourcing continues today, although the rationale has shifted from the geographic specificity of natural ingredients to the financial advantages of economies of scale and lower costs of production. As with almost all consumer products in the modern era, supply chains for pharmaceuticals are global; 72 percent of the manufacturing facilities for active pharmaceutical ingredients consumed in the US are in foreign countries.[3]

Adulteration Becomes a Crime

Pharmaceuticals are a natural but unfortunate target for fraud by dilution or substitution because of the high cost of exotic raw materials, the high value of cures for dreaded diseases, and the difficulty consumers have in verifying the substances in a drug. Moreover, the potential health consequences of substandard, contaminated, or non-sterile products make negligent manufacturing of drugs a public health risk. In 1540, the Apothecaries Wares, Drugs and Stuffs Act in England, which empowered inspectors to supervise drug manufacturing, began a slow but steady trend toward regulating pharmaceuticals.[4]

Today, government regulators oversee most aspects of pharmaceutical product invention, production, distribution, and marketing. They govern their manufacture through an extensive set of guidelines and requirements known as GMP[5] or CGMP (current good manufacturing practices). Although regulators from different governments may differ in their regulatory specifics,[6] they generally seek to ensure that every manufactured and administered dose of the product has the same safe and effective properties as the clinically tested and approved substance. That also implies showing that the product is still safe and effective across the range of proposed storage conditions, shelf-life recommendations, and logistical handling processes.

GMP is also a supply chain issue in that every raw material and supplied ingredient must also adhere to the regulations of the nation where the end product using it is ultimately sold—regardless of source. Thus, any foreign supplier providing raw materials or ingredients that go into a pharmaceutical used in the US is subject to US GMP regulations.

The Devil in the Details

If the fast invention of successful vaccines needed large numbers of scientists, each contributing key bits of knowledge to the bigger picture, the fast production of billions of doses of vaccines needed large numbers of supply chain professionals and engineers to create all the manufacturing systems and logistical processes required to supply raw materials, intermediate products, and the final products in the right volumes at the right time. "Probably the biggest challenge will be scaling up the actual vaccine," said George Zorich, a pharmacy expert and CEO of pharmaceutical consulting firm ZEDpharma. "It's one thing to have clinical trial samples and materials in lab quantities. It's another challenge actually scaling that up effectively."[7]

As mentioned in Chapter 1 (p. 19), Covid-19 vaccine manufacturing efforts began, quite unusually, in parallel with vaccine development. A person directly involved with the development of Pfizer's vaccine told the *Wall Street Journal*, "We started setting up the supply chain in March, while the vaccine was still being developed. That's totally unprecedented."[8]

Large-scale, global product launches create a nested double challenge for supply chains in terms of managing the ramp-up of suppliers and the production, packaging, and distribution of a new product. That double challenge can be especially difficult for unique and innovative products. First, launching such a product means creating new, end-to-end supply chains and manufacturing capacity for all the novel raw materials, parts, as well as the distribution of the finished product. Second, a new product launch then requires filling the inbound pipeline with synchronized flows of the ingredients and stocking the outbound distribution channels with the finished product.

As Moderna, Pfizer, Johnson & Johnson, and many others started their efforts to develop, test, produce, and distribute Covid vaccines, they ran into bottlenecks. Many of the highly specialized supply chains needed for a vaccine's journey did not have the scale needed to support the entire world's response to the pandemic. These bottlenecks appeared from the very beginning as vaccine developers searched for a viable vaccine, and they extended all the way to the sites that administered the shots.

Will the Lab Delay the Jab?

All of the vaccine makers, even those for the new mRNA ones, had decades of science under their belts and had confidence that their vaccine could help the world defeat Covid. However, the history of vaccine development has shown that the combined complexities of the human body and a new pathogen mean that every new vaccine has some unknowns in terms of both safety and efficacy. As described earlier, vaccines, like all medical products, require a rigorous program of preclinical and clinical testing. Those preclinical and clinical tests call for their own supply chains of testing-related materials, which spawned the first bottlenecks in the race to vaccinate the world.

Lab Supplies in Short Supply

Similar to Covid's effects on consumer supply chains,[9] the pandemic caused disruptions and shortages in medical research supply chains[10] as more than 100 research groups raced to develop vaccines and as many other research groups sought treatments for Covid. Moreover, the overlaps between the needs of researchers and those of the labs that were administering millions of Covid tests put extreme stress on laboratory supply chains. As with other supply chains, Covid-related factory shutdowns, border closures, and disruptions in air freight affected researchers' abilities to procure supplies such as pipettes, centrifuge tubes, and gloves.[11] The FDA warned that these shortages would last for the "duration of the Covid-19 Public Health Emergency."[12]

Researchers and clinicians found various tricks to deal with these issues. For one, researchers began sharing with other researchers and swapping their surplus supplies for the ones they needed. They found ways to clean and reuse some single-use products when it was safe to do so. For example, lab workers sequenced their use and reuse of personal protective equipment (PPE): They would put on fresh gloves to protect the integrity of the results during phases of the work that required sterility, and then they would switch to an old pair of gloves for non-sterile work where only the researcher's hands needed protection.

Monkey Business

Lab animals play a crucial role in the early stages of vaccine research. Despite strenuous objections from animal rights groups, the similarity of monkeys' pulmonary and immune systems to those of humans have made monkeys integral to the later preclinical stages of studying treatments and vaccines for diseases like Covid. In particular, monkeys can be used in challenge trials: intentionally exposing the animal to the virus and intensively monitoring its immune response and any side effects. Other, less controversial lab animals, such as mice, hamsters, and ferrets, have less biological similarity with humans, which gives scientists less confidence that results observed in these other animals reflect the true safety and efficacy of a vaccine or treatment on humans. When Covid arrived, however, it simultaneously increased demand and knocked out supply.[13]

The potential role of live animals at the Wuhan wet market in fostering the spread of Covid prompted Chinese authorities to immediately ban sales of many kinds of animals and prohibit their export. At the time of the ban, Chinese suppliers provided about 80 percent of the monkeys used in US research.[14] Even Chinese researchers were cut off until they could work through the stringent, multi-agency approval processes enacted by the Chinese government. Furthermore, despite China's shutdown of exports and regulatory obstacles, even Chinese researchers' demand for monkeys outstripped the country's supply.[15]

This shortage of lab animals was not unexpected. A 2018 NIH report noted a lack of "surge capability for unpredictable disease outbreaks" for non-human primates.[16] But, as with so many other ignored warnings about potential pandemic risks, the US had not yet addressed the issue. Adding to the problem was the refusal of major airlines, including the major American carriers—American, United, and Delta—to transport animals used in medical research, bowing to pressure from animal rights groups.[17] Moreover, lab animal supply chains are at the mercy of biology; monkeys reproduce very slowly and take years to mature.[18]

To deal with the shortage created by Covid, the NIH formed a pre-clinical working group to prioritize and optimize the allocation of animals and collaborate on solutions. For example, some research teams collaborated to harmonize their research protocols so that the teams could share one group of untreated control animals rather than each team needing its own control group. "It's really impressive," said Jeffrey Roberts of the California National Primate Research Center. "I've been involved in non-human primate research for 37 years, and I've never, ever seen this degree of coordination between different research institutes."[19]

The Patient as the Factory of Tomorrow

Manufacturing a vaccine is all about manufacturing an antigen, which is some part or analog of the pathogen that a person's immune system will recognize as a foreign substance—and thereby learn to attack the actual pathogen. Like preparing a complex recipe for a signature dish at a Michelin three-star restaurant, manufacturing a pharmaceutical like the antigen in a vaccine entails a long sequence of ingredients, with many steps for preparing, mixing, heating, cooling, and separating the intermediate products to make the desired final product. In most pharmaceutical factories, shiny stainless-steel vats sit amidst a spaghetti tangle of pipes, pumps, valves, and sensors. Control boards or computer screens choreograph the flows of fluids, the introduction of the next ingredient, the heating or cooling of mixtures, the filtration or separation of intermediate products, the pumping out of the final product into a storage vessel, and then feeding the bulk product out into an assembly line for packaging the drug into small vials, capsules, or pills.

Using the Tricks of the Oldest Chef in the World

Humans have perhaps a million or so years of experience with cooking and chemistry, dating back to the first campfires. That knowledge has enabled people to manufacture hundreds of thousands of different chemicals for industrial uses. However, the recipe for the SARS-CoV-2 spike protein—the antigen used by many vaccines—is orders of magnitude more complex than those of most industrial chemicals: It calls for connecting an exact sequence of 1,273 amino acids into a long chain.

Omit or misplace just one bead in the chain, and the resulting protein may not look anything like the spike protein or produce the required immune system response.

In contrast, biology has some 3 billion years of experience with cooking up intricate biological chemicals, such as proteins, RNA, DNA. In 1931, pathologist Ernest Goodpasture invented the method of growing viruses for vaccines inside chicken eggs. The egg is a natural, easily manufactured, self-contained factory for mass-producing many types of viruses. Goodpasture's method enabled the development of manufacturing systems for vaccines for influenza, chicken pox, smallpox, yellow fever, typhus, and other diseases.[20] The process involves injecting a small sample of the target virus into fertilized hens' eggs and incubating them for several days to allow the virus to replicate. The fluid containing the mass-produced virus is then harvested out of the eggs, isolated, inactivated, and purified. Finally, it is added to the formulation that became the vaccine.

While eggs are still important for growing viruses for the production of some vaccines, the egg-based manufacturing process is vulnerable to supply chain disruptions. These problems can be rooted in bad weather, contaminated eggs, hens' nutritional deficiencies, poultry diseases, and so forth.

From Eggs to Tanks to Arms

More recently,* scientists and manufacturers found ways of growing a batch of viruses in cell cultures in large, stainless-steel tanks.[21] These bioreactors or fermenters can culture thousands of liters of cells in a warm, delicious nutrient broth, with each cell in the fermenter being its own tiny factory for both more cells and the desired product, such as the antigen for a vaccine. When the cells have completed their growth cycle and made the product, the factory operators decant the tank, separate the cells from the product, and further process or purify the product as needed.

The development of genetic vaccines, such as those based on mRNA and viral vectors, take this biological factory trick one step further. Genetic vaccine technologies use the patient's own cells to mass-produce

---

* A cell-based production process for flu vaccines was approved by the FDA in 2012. In 2016, the FDA approved Flucelvax, a cell-based flu vaccine made by Seqirus, based on cell-grown viruses.

the required antigen—e.g., Covid's spike protein—that then trains the immune system to recognize the virus. This trick saves on the delays of manufacturing the antigen (i.e., quantities of the protein or inactivated viral particles needed to trigger an immune response). Moreover, in mimicking the infectious process in which a virus hijacks human cells to make foreign proteins, these vaccines also train the immune system to spot and kill infected cells. "Traditional vaccine manufacturing methods... are not vertically scalable," said Dr. Parviz Shamlou, head of the Jefferson Institute for Bioprocessing at Thomas Jefferson University. "Advanced vaccines are vertically scalable[†] and have the potential to become the basis for a trillion-dollar vaccine industry as we come out of Covid-19."[22]

Recipe for an mRNA Vaccine: Serves 1,000,000

Manufacturing of mRNA uses refined enzymatic processes that borrow the minimum set of nature's biological enzymes and biochemical ingredients required for replicating genetic strings. This avoids the challenges—and delays—of the care, feeding, and eventual removal of complete cells in a fermenter. "Traditional biotechnology requires these massive bioreactors, because there's a limited space into which you can crowd cells before they get really, really unhappy. But enzymatic reactions actually like to be dense—in fact, they become more efficient," said Moderna's chief medical officer Tal Zaks at Biotech Week Boston in 2020.[23]

The minuscule dosage of an mRNA vaccine (100 micrograms in the case of Moderna) means that a million doses of Moderna's vaccine require only a few ounces (100 grams) of mRNA. A batch with a million doses can be made in a container no taller than the average person. That batch can be brewed in a container of a few hundred liters containing a cocktail composed of the DNA template for the vaccine, some enzymes, and the building-block-modified nucleosides for the mRNA.

[†] *Vertical scalability* means that more resources can be quickly added to an operation and it can perform on a larger scale without changing the underlying process or incurring substantive delays.

"If you understand the fact that it's a simple enzymatic process at the end, and the scale is relatively small, it explains why we've been both [Moderna and Pfizer] able to start our trial so fast," Zaks added.

Gathering the Ingredients

As Moderna's first clinical trials of its Covid vaccine progressed in spring of 2020, Paul Granadillo, Moderna's senior vice president of supply chain, realized, "Okay, we're going after something at scale."[24] That scale was far larger than anything previously done at Moderna or the entire segment of the biotech industry devoted to mRNA therapies. Now multiple vaccine makers were looking to make unprecedented amounts of mRNA, accompanying vaccine ingredients, and excipients (inactive ingredients needed to package and stabilize the mRNA). "We started going from talking in milligrams or grams to talking in grams and kilograms. It was big shifts," Granadillo said. "I would say that the first chapter of the journey for me and my team really had everything to do with materials. We weren't quite sure on demand, we weren't quite sure what the final scales would be, but we knew we needed a lot of materials."

As the enormity of potentially trying to vaccinate billions of people as quickly as possible sank in, Moderna's supply chain team asked themselves, in the words of Granadillo, "What do we need? How do we go get that? Can our suppliers do that today? If they can't, who else can try to help us? If they can't, what CMOs [contract manufacturing organizations] can they potentially work with to expand their capacities?" This made the company very concerned about potential production ceilings for mRNA and the required ingredients. "There were several times—and I'll point to lipids and enzymes alone—where we said, 'Gosh, I don't see how we get over this ceiling.'"[25]

Moderna worked closely with its suppliers to try to bust through each new ceiling. "It was time after time of finding a way to continue forward and reach new scale," Granadillo said. "Through working with suppliers and suppliers going back recognizing the severity of what we're asking for, they found a way. People were very much willing to work together between government, suppliers, innovators, and CMOs in collaborative ways that we typically wouldn't have seen," Granadillo continued. Agreements that probably would have taken months or years to negotiate were taking days or weeks to complete. "I think it's a testament to the goodness of people," he concluded.

A molecule like *heptadecan-9-yl 8-((2-hydroxyethyl)(6-oxo-6-(undecyloxy)hexyl)amino)octanoate* sounds like an extra-credit question on a graduate-level organic chemistry final exam. In reality, this ionizable lipid molecule—called SM-102—is a key part of the four-lipid mixture that forms the nanoscopic bubble of packaging that helps ferry the fragile mRNA molecules into the cells of the vaccine recipient.[26] SM-102 is almost like the "tear here" sealing strip on an envelope that helps safely seal the mRNA inside the lipid bubble and then easily open up once it gets in the patient's cells. Making SM-102 requires a careful sequence of about a dozen steps to synthesize its multiple branching chains of exactly the right substructures and lengths.[27] That synthesis is followed by a careful purification process that requires highly specialized equipment.

CordenPharma, a German contract development and manufacturing multinational, has such equipment and helps make and purify the lipid excipients used in Moderna's vaccine.[28] Matthieu Giraud, director of peptides, lipids, and carbohydrates at CordenPharma, said that the company had completed ramp-up preparations in anticipation of the high-volume vaccine production.[29]

"It hasn't been easy," Giraud said, regarding the effort to boost capacity more than tenfold in a matter of months. "We had to leverage our whole network." Andrey Zarur, co-founder of Boston's Greenlight Biosciences, another company working on RNA-based products, said, "I don't want to give you the impression that once we solve the lipid nanoparticle issue, then the 16 billion doses for humanity are solved. Because the reality is, we solve that bottleneck, and then we'll find another bottleneck."[30]

The Latest in Tiny Tech to Make the Latest in Tiny Particles

A unique, key step in making an mRNA vaccine is merging the mRNA strands with the encapsulating lipids to form huge numbers of tiny lipid nanoparticles, each containing a strand of mRNA. This requires molecular-level mixing of the mRNA (suspended in water) with a concoction of the lipids (dissolved in ethanol). The manufacturing process for this step highlights the key role of other advances in technology—outside of biotechnology—over the decades leading up to the Covid pandemic.

Microfluidics is a fluid-handling technology in which fluids (liquids or gases) are manipulated inside a device containing tiny channels,

chambers, and other features.[31] Microfluidics borrowed the manufacturing techniques of the semiconductor industry to quickly and inex-
pensively fabricate the required intricate, miniaturized plumbing on
a flat plate.[32] In essence, microfluidics is a way of mass-producing a
tiny fluid-handling factory suitable for either working with very small
quantities of substances for laboratories or performing very specific
chemical, biochemical, or fluid-manipulating operations.

Pfizer uses a microfluidics plate about the size of an Apple Watch
face (roughly 1.5 inches or 40 mm) to manufacture its mRNA vaccines.
Called an impingement jet mixer, it has two streams of raw materials
that enter the plate at high pressure (450 pounds per square inch) and
are forced into channels about the size of a human hair. The streams
collide, mix, and form large numbers of nearly uniform bubbles of lipid,
most of which contain the essential mRNA ingredient.[33] Large-scale
versions of this type of mixer had never been made before, and Pfizer's
engineers knew they did not have time to design, build, test, and deploy
an industrial-scale version of the mixer. Instead, they stacked multiple microfluidic plates and used eight pairs of carefully synchronized
pumps to drive more fluids through the parallel assembly.[34]

The resulting mix then goes into a special tangential-flow filter
that separates the precious mRNA-containing nanoparticles from the
remaining unused lipids and liquid raw materials. Pfizer's engineers
first tested their new mixer-filter system in mid-September 2020. "Our
first engineering trial," said Pat McEvoy, Pfizer's senior director of operations and engineering, "was an absolute and utter failure."[35] Analysis
showed that the filter had failed, letting the product escape into the
waste stream. Although Pfizer's engineers found a quick fix and had
success three days later, these filters continued to be a challenge. To
increase production of their vaccine, they found they needed larger filters. To deal with shortages of filters, they developed methods for cleaning and reusing filters.

Flexible Space for Fast-Built Factories

Vaccine production depends on having the right physical facilities,
specific manufacturing equipment capacity, and systems to make the
product. Although Moderna knew from its founding in 2010 that development of new pharmaceutical products could take a decade or more,
the company also knew that a viable product could appear faster than
expected. Commercial products were Moderna's ultimate goal, and the

company wanted to have its own large-scale manufacturing facility even though small-scale lab work had dominated its early years. That goal and the realities of pharmaceutical development influenced the company's facility designs.

"We needed to be planning this facility like a commercial industrialized manufacturing facility and not clinical lab space," said Moderna's Paul Granadillo.[36] Having a disciplined, digital approach to scheduling and managing low-volume clinical activities meant that the facility had the digital processes required for high-volume production. Moreover, the company designed its manufacturing white room for flexibility by having an open space with utilities suspended from the ceiling to make it easy to rearrange the space.

For manufacturing, Moderna chose a strategy based on single-use equipment, such as bioreactor bags, plastic tubing, and connectors, rather than traditional fixed, stainless-steel equipment. "Part of the reason we are leveraging so much single-use technology," Granadillo explained, "is because of the speed that it allows us to proceed forward with, and the flexibility to change the assemblies." Single-use equipment also avoids the costs, delays, and risks of cleaning protocols.[37] Although using single-use plastic bags in manufacturing seems, at first glance, to be wasteful, the disposable bags actually have a lower environmental footprint because using them eliminates all the chemicals and hot water required to clean a reusable stainless-steel vessel after each use.[38] Moderna uses suppliers who can build presterilized assemblies of bags, tubing, connectors, filters, and so on to whatever design specifications Moderna needs for the task at hand. The result is maximum flexibility, maximum production speed, and a minimum risk of contamination.

Vaccine Factory: Some Assembly Required

Even for the giants among global pharmaceutical companies, it was a significant challenge to design, build, test, and operate a manufacturing system for an entirely new type of product on an accelerated schedule. Chaz Calitri, Pfizer's vice president of operations for injectable drugs in the United States and Europe, described the stakes: "The weight of the world was on us. We have the manufacturing capability for a solution to the pandemic, and we knew we couldn't go fast enough."[39]

Although Pfizer had never manufactured an mRNA vaccine in high volume—no one had—most of the steps for making BioNTech's mRNA

vaccine overlapped with Pfizer's vast array of pharmaceutical manufacturing capacity for making biologic pharmaceuticals.[‡] "We built this
out of the Erector Set we had," said McEvoy, who oversees the sprawling
1,300-acre plant in Kalamazoo, Michigan, where much of Pfizer's US
vaccine manufacturing would take place.[40]

Like Moderna, Pfizer also used a prefabricated, modular strategy for
production equipment. At Pfizer, these were entire rooms pre-built by
a contractor in Texas, trucked to Kalamazoo, and rolled into place. "We
had planned to expand our formulation capacity," said Mike McDer-
mott, Pfizer's president of global supply. "The question was, how can we
do it quickly? If we built it wall-by-wall on-site, it would have taken us
a year. By doing it modularly, we could cut that in half."[41]

Bottling the Batches

In the end, a single batch of mRNA weighing only a few ounces is enough
for a million doses. That batch expands to a volume of 500 liters (132
gallons) once it is processed into lipid nanoparticles, diluted with water
(to create an easy-to-inject amount), buffered (to stabilize the pH of the
vaccine), and mixed with a bit of sugar (to protect the lipid nanoparticles during freezing). Fill-finish is the final step, in which bulk quantities of vaccine are subdivided, put in vials, capped, inspected, labeled,
packed, and put into temporary storage (freezers in the case of mRNA
vaccines).

Moderna signed deals with several companies (including Catal-
ent, Rovi, and Recipharm) for fill-finish.[42] This multi-partner strategy
reflected Moderna's need to cobble together capacity to support both
high volumes and worldwide distribution. It also reflected the firm's
strategy to focus on its core strengths and contract out generic manufacturing processes.

What a Pain in the Glass

The next production bottleneck was essentially caused by bottles: the
10-cent glass vials used to package some modest number of doses of
a vaccine for injections into people. The first warnings of a potential

---

[‡] Biologic pharmaceuticals are typically synthetic proteins made using enzymatic or cell culture techniques. They contrast with so-called small-molecule
drugs made using standard chemistry techniques.

shortage occurred in early March 2020.[43] Vaccine and drug makers need a small, robust container that is chemically resistant, doesn't affect the vaccines, prevents oxygen from diffusing into the container, and can handle the shocks of sterilization, freezing, and handling. "Not all glass is the same," said Howard Sklamberg, a former FDA official. "A manufacturer making champagne glasses can't just switch to making medical vials."[44]

Most supply chain disruptions involve shortages of parts or material needed to make products. When shortages happen, the affected suppliers can use any of several possible allocation strategies for handling product shortages so as to satisfy as many customers are possible without losing customers' long-term good will.[45] During the pandemic, however, the largest supplier of borosilicate glass for vials, Schott AG, used an unusual allocation strategy when the big pharmaceutical companies tried to lock in supplies of vials. Schott declined preorders of some 800 million to 1 billion vials until the vaccine maker could prove it had a viable product. Schott's chief executive, Frank Heinricht, told Reuters, "We have to keep the door open to give capacity to those who really are successful in the end."[46]

"No one wants to see glass be the reason the world can't get access," said Brendan Mosher, vice president and general manager of Corning Pharmaceutical Technologies. "Everyone is working together and has a common goal to make sure there are plenty of vials."[47] As part of Operation Warp Speed, the US government paid companies such as Corning and SiO2 Material Science to expand capacity for medical glass and vials.

Others found ways to extend vial supplies. Moderna obtained approval from the FDA to fill vials with 14 doses instead of the original 10.[48] Those handling Pfizer's vaccine found they could get six doses out of a five-dose vial if they used syringes with a low dead volume.[49] That practice, now widespread, both reduced the number of vials and extended supplies of the vaccine.

The potential to boost Pfizer's vaccine supplies by 20 percent through the use of low-dead-volume syringes caused a surge in demand for those syringes, resulting in other shortages. One day after his inauguration, US President Joe Biden invoked the Defense Production Act (DPA), by which the government can essentially compel businesses to manufacture certain products, to deal with shortages of these syringes and eleven other critical categories of supplies.[50] That executive order also addressed other constraints in Pfizer's supply chain, such as filling pumps and tangential flow filtration components. The White House's supply coordinator, Tim Manning, said the order "will allow Pfizer to ramp up production and hit their targets of delivering hundreds of millions of doses over the coming months."[51]

Sourcing from Near or Far in a Time of Crisis

As Covid raced around the world, it disrupted global supply chains when facilities, countries, and transportation systems locked down, imposed social-distancing constraints, or experienced extreme changes in demand. For example, before Covid, passenger aircraft were providing just over half of the world's airfreight capacity (in "belly cargo"), including 45 percent of Asian airfreight capacity and some 80 percent of transatlantic capacity.[52] The steep reduction in passenger flights at the time when the vaccine makers were ramping up took most of this capacity out of the market. For vaccine makers and their high-value, time-sensitive shipments such as raw materials, test samples, and finished vaccines, the pandemic's impact on airfreight was especially challenging. Johnson & Johnson's Meri Stevens explained: "Prior to the pandemic, about 70 percent of our J&J medical and pharmaceutical products were transported in the bellies of commercial aircraft."[53] The ripples of disruption caused some to criticize global trade and to advocate more localized networks of supply and production. Moreover, political forces during the global crisis also favored self-preservation and self-reliance (vaccine nationalism is discussed in Chapter 3, p. 64).

"The push to buy local-for-local and to shorten supply lines was not something new from the pandemic," Stevens said. However, the long regulatory approval times required to certify a new facility for a supplier meant that pharmaceutical companies like J&J had to avoid knee-jerk responses when investing in manufacturing capacity. "We're not rushing to say, 'Now we're going to move everything from A to B,' because that's expensive and that takes time," Stevens said. In the case of J&J, she continued, "During the pandemic, what we found is that the global nature of our supply chains worked really well—the balance that we have around the world was very good. While we have had volatile demand, which has caused localized shortages, we actually had very few critical product production stoppages other than a few odd constraints, like alcohol when everybody was making hand sanitizer and wipes."[54]

As vaccine makers formulated their vaccines, they also formulated their sourcing and manufacturing site selection strategies for vaccine production. That raised trade-off questions between large-scale centralized production to serve global demand versus decentralized production with sites serving local demand. At J&J, the supply chain design criteria emphasized reliability and resilience to serve customers and patients in the most effective and efficient ways. That said, Stevens also noted, "We'll also continuously evaluate the national norms that are going on to see whether or not we have to rebalance."[55]

Pumping Up Production

Vaccinating even 70 percent of the world's population of 7.8 billion would require nearly 11 billion doses of vaccine (assuming two shots per person, which is a common requirement of many of the leading vaccines). Moreover, the potential need for booster shots to address either natural declines in immunity or the emergence of variants of the SARS-CoV-2 virus (see Chapter 5, p. 104) brings a high likelihood of demand for additional billions of doses in the future. Although a number of leading vaccine producers were targeting volumes of billions of doses a year, the scale of the challenge of vaccinating the world as fast as possible meant a clear need for much higher global capacity. In the quest to vaccinate the world, companies sought partners, innovated new ways of building factories, digitalized operations, and optimized production processes.

Finding a Big Buddy

While BioNTech had its mRNA vaccine technology, the small German biotech startup knew that getting to scale quickly would require access to significant resources. In March 2020, they chose to partner with Pfizer (the second-largest pharmaceutical company in the world) with whom they already had a 2018 agreement related to developing mRNA-based influenza vaccines.[56] (Pfizer revenues in 2019[57] were almost 500 times that of BioNTech.[58]) In announcing the partnership, Mikael Dolsten, chief scientific officer at Pfizer, said, "We believe that by pairing Pfizer's development, regulatory, and commercial capabilities with BioNTech's mRNA vaccine technology and expertise as one of the industry leaders, we are reinforcing our commitment to do everything we can to combat this escalating pandemic, as quickly as possible."[59]

Pfizer poured $2 billion of its own money into the aggressive project. "For this one, everything happened simultaneously," a person familiar with Pfizer's efforts told the *Wall Street Journal*. The vaccine's manufacturing and supply chain were created while the product was still under development.[60] Mike McDermott, Pfizer's president of global supply, added, "My team spent $500 million, before we even got out of clinical trials. So, all completely at risk. We didn't know if we had a product that was going to work."[61]

More Partners = More Production

Many of the vaccine makers sought partners in order to quickly access needed capacity and geographic diversity. "We're leaving no stone unturned in terms of partnerships," said Alex Gorsky, J&J's CEO. "One of the most important lessons of the pandemic is the power of collaboration."[62] Paul Lefebvre, who heads the Covid-19 vaccine supply chain at Janssen Pharmaceuticals[§] said, "In addition to establishing a capable manufacturing network, it's really important to have multiple manufacturing sites located in various regions to maintain the continuity of our supply chain."[63] J&J's Stevens concluded, "Really, it is about, how do we think about partnerships? How do we think about leveraging sources of supply? But [it's] also really [about] continuously balancing our supply lines to serve the world in the most effective way."[64]

Other vaccine makers also sought partners. For example, Moderna contracted with Lonza, a Swiss pharmaceutical contract manufacturer, to further increase the scale of production, especially in Europe. That contract manufacturer too had been preparing for rapid scaling. Two years earlier, Lonza invested in a shell-based strategy for building new factories. Instead of constructing purpose-built facilities for specific products after Lonza got a contract, the company designed and pre-built empty shell buildings that were pre-equipped with all the basic utilities and facilities (sterile water, steam, gas, data networks, etc.) needed for pharmaceutical manufacturing. Torsten Schmidt, leading the production facility in Switzerland, said, "The empty shells allow us to drop in the manufacturing technology that is needed for a particular drug or vaccine. This is important given that vaccines and drugs are

---

[§] Janssen Pharmaceuticals is the Belgium-headquartered, wholly-owned subsidiary of J&J that developed, manufactured, and distributed the J&J Covid vaccine.

becoming more diverse and a facility for one drug or vaccine cannot be easily used for another type of molecule."[65]

Andre Goerke, Lonza's global lead for the Moderna project, said in an email to the American Chemical Society's *Chemical & Engineering News*, "Since we signed the agreement with Moderna in May this year [2020], the focus has been on getting four manufacturing kits up and running, each capable of producing an estimated 100 million doses of mRNA-1273 per year."[66] Schmidt added, "In this pandemic situation, we are working around the clock to set up manufacturing in around eight months, compared to the two or more years it would usually take."[67]

Accelerating the Learning Curve for Production

Vaccine manufacturing operates at different scales as it progresses from lab to jab: research scale (the equivalent of a few doses for *in vitro* or *in vivo* animal studies), clinical trial scale (hundreds to tens of thousands of doses), and mass production (millions to billions of doses). At each stage, the vaccine maker's scientists learn something about the effectiveness and safety of the vaccine. Similarly, at each stage, the vaccine maker's production engineers learn something about the speed, yield, and processing parameters of making the vaccine. Getting to scale quickly means being able to learn and apply lessons quickly, and that implies being able to make the best use of data.

Along those lines, to accelerate development and manufacturing at Moderna, Marcello Damiani, the company's chief digital and operational excellence officer, said, "We decided from the get-go to make it paperless."[68] Moderna's all-digital strategy starts in the research phase. Researchers use a web platform for designing genetic sequences for new mRNA products, which enables direct production of research sequences in a fully automated central lab. Internal digital collaboration between scientists and engineers helps improve those genetic sequences for manufacturability and yield.

All of Moderna's equipment connects into its digital platform so that every step of the process collects data. "Between the small-scale research to the large-scale manufacturing, if you collect data from the different type of instruments, you are learning, and that's the key piece," Damiani added. "Once you have the automation, the Internet of Things and the integration on the Cloud, you have data that's flowing, and you can start doing sophisticated analytics," he continued.[69] "All this learning started at very small scales, and with this learning we had in place, we built our clinical manufacturing and GMP manufacturing." Damiani concluded by saying: "So, you see how we built the company, and I think the line between all this is data, data, data, because we collect the data to improve."[70]

The Quest for Quantity: From Setback to Upsurge

In early fall 2020, Pfizer's chairman and chief executive Albert Bourla told employees, "Every ounce of our ability has been spent and nearly $2 billion put at risk."[71] Initially, Pfizer had hoped to deliver 100 million doses by the end of 2020 and 1.3 billion doses in 2021.[72] Despite the all-out effort, Pfizer faced serious challenges in production and supply chain operations, as scaling up the raw material supply chain took longer than expected.[73] In November, the company realized it could not achieve its original production targets and cut the 2020 production forecast in half, to 50 million doses.

As Pfizer and BioNTech worked to overcome the obstacles to fulfilling their original forecasts, they also began pushing to vastly exceed those forecasts. They expanded their European manufacturing network from three partners to 13.[74] BioNTech also recruited more manufacturing capacity from other larger pharmaceutical companies, such as Novartis and Sanofi.[75] BioNTech reported capacity-boosting initiatives that included "the optimization of production processes, the recent initiation of production at BioNTech's Marburg, Germany facility, regulatory approval for six-dose vials, and the expansion of [its] manufacturing and supplier network."[76] Manufacturing engineers cut the batch production time nearly in half (from 110 days to about 60). Improvements in uniformity cut the waste on vial inspection lines from about 5 percent to about 1–2 percent.[77]

Some of the efforts to boost production caused short-term delays in deliveries, such as when Pfizer renovated its Belgium facility in January to increase its capacity.[78] During the upgrade, Pfizer suspended vaccine deliveries to Europe and Canada. European authorities threatened legal action because the delays forced them to suspend or reduce vaccinations.[79] Charles Michel, president of the European Council, told radio station Europe 1, "We plan to make the pharmaceutical companies respect the contracts they have signed... by using the legal means at our disposal."[80] In the end, the companies' efforts paid off, as Pfizer and BioNTech issued a steady stream of announcements during 2021 upping the 2021 delivery forecast to 2 billion doses in February,[81] 2.5 billion in March,[82] and 3 billion in May.[83]

The Vaccine that Came in from the Cold

"Ensuring over a billion people globally have access to our potential vaccine is as critical as developing the vaccine itself," said Pfizer's CEO Bourla.[84] Adding to the challenge of both the volume of shipments and the urgency of delivery was the need to properly handle the vials of vaccine while sending them to the far corners of the earth.

Whereas most vaccines require some refrigeration, the new mRNA vaccines require the most careful handling because of the delicate constitution of their lipid nanoparticles. Molecular biologist Phillip Sharp of MIT explained, "This is an oily particle with carbohydrate around it. So, it's a pain to keep it from fusing. It's just one big ball of oil if it's not taken care of. That's why all this shipping and freezing and thawing and everything is really very important."[85] Moderna's vaccine requires freezing between –50°C and –15°C (–58°F and 5°F), and Pfizer's requires ultra-low-temperature freezing between –80°C and –60°C (–112°F and –76°F). As a result, these vaccines require cold-chain handling: global distribution activities at very low and controlled temperatures.

The colder the temperature, the more challenging the cold-chain transportation and storage issues. In the case of the Pfizer vaccine's deep-freeze needs, very few facilities—only a handful of pharmaceutical distribution centers, hospitals, and research laboratories—had the kinds of deep freezers needed. "I don't think we have all the cold storage that people think we have," commented James Bruno, president of the consulting firm Chemical and Pharmaceutical Solutions.[86]

Helping Shipments Keep Their Cool

As part of its parallel development strategy, Pfizer began setting up its downstream supply chain for the finished product in March 2020—at the same time as the kick-off of its Covid vaccine development. Pfizer said it developed a "just-in-time system, which will ship the frozen vials direct to the point of vaccination."[87] That system included packaging for shipping, continuous monitoring of vaccine temperatures to ensure safety, and a means to store the vaccine for up to a month at clinics,

vaccination centers, and distribution facilities that lacked deep freezers. These efforts used supply chain partners with respective expertise
in cold-chain packaging and supply chain monitoring.

Pfizer worked with SoftBox, a multinational British manufacturer
of temperature-controlled packaging, to develop a reusable insulated
thermal shipping box[88] that holds 1,200 to 6,000 doses (at the 6-dose-
per-vial capacity). The box, measuring 17 × 17 × 22 inches, holds up to
five small "pizza box" trays, each with 195 vials, in an inner payload
sleeve box nestled deep in the heavily insulated outer box.[89] On top of
the precious cargo sits a "pod" with up to 50 pounds of dry ice at –109°F
(–79°C). An insulated lid completes the cozy ensemble.

The result is a medium-sized, robust, 70- to 80-pound box (with side
straps) that can be handled by any air or ground parcel delivery service.
(Pfizer and SoftBox even designed the box to reduce the sublimation of
the dry ice during flight, reducing the generation of potentially hazard-
ous CO_2 levels in air-freighters and significantly increasing the number
of doses that air-freighters were permitted to safely carry.)[90]

As an added bonus, this thermal container can maintain ultra-cold
temperatures for up to 10 days. Moreover, if needed, the recipient can
replenish the dry ice every five days to extend storage in the box for up
to 30 days.[91] That enables facilities that lack the required freezers to
temporarily store, distribute, and dispense the vaccine. Finally, when
needed, the vaccine is thawed and can be kept in an ordinary refriger-
ator for up to five days before dilution and injection.

A View to a Chill

To maintain 24/7 visibility onto shipments, Pfizer contracted with Controlant, a provider of real-time supply chain monitoring devices that go
into shipping boxes.[92] A small, battery-powered sensor tracks vaccine
temperature, the opening of the box, and its GPS location. Using a standard cellular data connection, the sensor sends the information in real
time to Controlant's cloud-based software, where customers can receive
alerts and view the information. When the box is opened, red-green
status lights show the shipment's temperature status, data connection
status, and battery status. "Controlant's reusable, real-time data log-
gers and visibility and analysis platform integrates with Pfizer's existing control tower technologies," said Tanya Alcorn, vice president of
biopharma global supply chain at Pfizer, "to help manage temperature

proactively, identify and react expeditiously to any events that can impact the supply chain, all while automating quality and logistics processes."[93]

One tricky issue with the monitoring system occurred at the hand-off when Pfizer delivered the doses to government distribution or vaccination centers. As the shipment left Pfizer's hands, Pfizer turned the monitoring off for legal liability and practicality reasons; once delivered, Pfizer had no control over the status of the shipment or the means to make the recipient take a corrective action. But recipients wanted the ability to monitor the boxes too, especially if they planned to use them for interim storage by refilling the dry ice. Fortunately, because the monitoring device was actually made and monitored by a third party, Controlant, all any recipient had to do was to sign up with the tracking company to restart monitoring and route the data and alerts to the recipient.[94]

Getting Ready to Move 'Em Out

Pfizer bought large numbers of deep freezers to set up freezer farms to buffer and distribute the output of its production facilities in Michigan and Belgium. The company also built its own dry ice plant to make the freezing pods that keep the vaccines cold during transit. As of November 2020, Pfizer planned to have a fleet of 24 trucks to ferry shipments from Pfizer's facilities to local airports, where a combination of air charter and air freight companies such as FedEx, UPS, and DHL could carry the vaccine anywhere in the world within a day or two. As the clinical results of the Phase 3 trials confirmed the efficacy of the Pfizer–BioNTech vaccine, Pfizer announced plans to move roughly 7.6 million doses per day.[95]

Similarly, airfreight companies and facilities prepared for the vaccine distribution campaign. UPS, for example, built its own freezer farms and dry-ice production equipment at key air hubs.[96] Airports invested in additional security and cold storage.[97] Airlines conducted trial runs of vaccine deliveries to both debug systems and ensure the CO_2 emissions from the dry ice remained within FAA-required limits.[98] In coordination with Operation Warp Speed, FedEx and UPS divided the US in half to improve delivery efficiencies.[99] The efforts were intended to ensure fast, efficient, and problem-free delivery of the vaccines once they were approved and started shipping.

The Bigger Picture of Bigger Demand

Overall, vaccine suppliers had to face and overcome a long list of challenges: Shortages began in the product development labs, moved into
the ingredient supply chains, and then hit the packaging ends of vaccine development and production processes. Shortages also hit capital
equipment supply chains as pharmaceutical makers attempted to ramp
up their capacity. As the adage goes, supply chains are only as strong
as their weakest links. Successfully delivering large quantities of a new
product depends on delivering all of the required quantities of every
one of the raw materials, ingredients, and all other parts in the bill of
materials (BOM) of the final product,[¶] as well as all the plant equipment
and machinery needed for manufacturing and delivering the product.
Supply chains aren't about doing one thing well; they are about doing
every one of many things well, because final products and customer
satisfaction depend on every one of those many things for a complete,
high-quality product delivered on time.

Even the packaged final product—billions of doses of safe and
effective vaccines—wasn't the end of the challenge. Those vaccines still
needed to get to the customers: all people in all the countries of the
world. Although modern supply chains have become adept at quickly
and accurately making and moving millions of shipments of consumer
products per day anywhere in the world, actually getting those doses
into people's arms was a real challenge that tested national and local
institutions.

[¶] A *bill of materials* is a nested list of the raw materials, sub-assemblies, intermediate assemblies, sub-components, parts, and the quantities of each needed
to manufacture an *end product*.

3. The Ins and Outs of Rollouts

A successful vaccination campaign, like a successful new product launch, requires a spectrum of supply chain activities, beginning with acquiring sufficient supplies and ending with successfully delivering the product—and, in many cases, the related service—to all the customers who need or want it. Semi-serious jokes about Amazon distributing vaccines ("We should train all Amazon delivery drivers to give the vaccine. The whole population would be immunized by Saturday. Thursday if you've got Prime.")[1] alluded to the online retailer's ability to deliver 300 million packages a month, with most arriving within a day or two of ordering.[2] However, getting a shot into someone's arm takes more than just putting a box on their doorstep. Vaccination requires a safe location for administering the injection, trained medical personnel, PPE, and careful coordination so that the right people get their vaccination at the right time.

New products often have a target market: the types of customers whom the seller of the product particularly wishes to reach. In the case of Covid vaccines, the disease's higher likelihood of severely sickening or killing the elderly, the obese, diabetics, and those with any of a wide range of medical conditions, made those groups high priorities for vaccination.[3] Moreover, given that no vaccine is 100 percent effective, especially in the elderly and those with immunocompromising medical conditions, vaccine providers also wished to reach people who often interacted with the most vulnerable populations, such as workers in nursing homes, hospitals, healthcare, front-line retail, and so on. For both individual consumers and the economy as a whole, the stakes for the effective rollout of vaccinations couldn't be higher: Every day's delay in delivering vaccinations meant thousands of preventable deaths, hundreds of thousands of new cases of Covid, and the ongoing economic impacts of social distancing and shutdowns.

Public health experts knew that the rollout of the Covid vaccines would pose many challenges. The product was in great demand by a large fraction of the population, necessitating a large-scale delivery organization. The product was in short supply—requiring allocation, control of who gets served and when, and strict minimization of waste. The product was delicate, requiring cold-chain storage and careful handling under sterile conditions.

The product's delivery required coordinated services (i.e., the installation of the product, in this case an injection), meaning setting up appointments and synchronizing the numbers of customers, vaccine doses, injection supplies, and workers at vaccination centers. Moreover, most of the vaccines also required two doses spaced at a particular interval, complicating both the allocation of supplies and coordination of delivery. For instance, if a location receives 10,000 vaccine doses, should it use all 10,000 immediately and hope that another 10,000 doses come in time for the second shot, or should it only give out 5,000 shots and hold the rest for the second dose?

"There's going to be a lot of tripping and falling," said Dr. Paul Offit, director of the Vaccine Education Center at the Children's Hospital of Philadelphia, in November 2020. "We're going to learn a lot over the next few months about how we probably could have done this differently."[4] The first months of the rollout provided these lessons: Some countries (e.g., Israel) did unexpectedly better than others (e.g., the US, UK, and EU). In the 100 days since approving the Pfizer vaccine on December 10, 2020, Israel fully vaccinated more than 52 percent of its population. By contrast, the US achieved a meager 13 percent and the EU merely 4 percent.[5] Some of those early laggards, however, managed to learn and improve.

Reaching the Promised Land of Vaccination

Like many other countries, Israel was striving to vaccinate as many of its citizens as possible as fast as possible. That drive began with acquiring sufficient supplies. Israel's Prime Minister Benjamin Netanyahu spent several months of the pandemic talking to CEOs of pharmaceutical companies, buying, cajoling, and using personal connections and non-economic incentives to ensure vaccine supplies for his people. Similar to the US, Israel made early commitments to acquire a sufficient number of doses, well before the country or the vaccine makers knew whether the vaccines would prove safe and effective.

Buy Early, Buy Often

Israel signed a deal with Moderna in mid-June, weeks before Moderna had even started its Phase 3 clinical trials that would finally show if the company's vaccine really worked.[6] Israel also hedged its bets: In November 2020, Netanyahu announced, "I reported today in the coronavirus cabinet meeting that I spoke again last night with Pfizer CEO Albert Bourla. Together with the legal advisers of both parties, we removed the last obstacle to signing a contract with Pfizer."[7] Israel received shipments of Pfizer's vaccine before the Pfizer vaccine was officially approved by the FDA.

"The principle is simple: Buy as many options as possible from as many companies as possible," Netanyahu told his cabinet. "The cost of buying vaccines is negligible, compared to the cost of not bringing them. The cost of throwing money in the trash, if the vaccines are unsuccessful, is minimal compared to not having vaccines."[8] At that level, Israel's strategy was no different than that of the US and the EU in risking billions up front to support vaccine development with the implicit or explicit expectation that the up-front support of vaccine development would secure early access to hundreds of millions of doses if and when those government-funded efforts bore fruit.

Although the cost in human lives is of paramount importance, the economic costs of not having a vaccine are staggering. Indeed, the huge imbalance between the costs of Covid—both social and economic—compared to the modest price of vaccines also meant Israel was willing to pay a few dollars more per dose to ensure timely shipments.[9] Israel paid $23.50 per dose, the US $19.50, and the EU $14.76. The pandemic had knocked about 6 percent off Israel's expected 2021 GDP,[10] amounting to an economic loss of some $240 per Israeli per month of ongoing restricted economic activity. The economic benefits alone—faster vaccination, earlier herd immunity, and an expedited reopening of the economy—easily justified the pittance of a premium the country paid. Naturally, the value of saving lives and averting serious illness even further justified paying more.

By comparison, the European Union bungled a key element of its vaccine purchase negotiations. On the one hand, the EU's good decision to centralize both the R&D funding and its procurement efforts, as well as to negotiate for the entire bloc, made sense and was praised by European Commission President Ursula von der Leyen. However, the EU then focused its attention on reducing the price of the vaccine rather than maximizing its value. EU negotiators haggled with the manufacturers on indemnification issues, trying to protect themselves from consequent criticism should something go wrong or from any accusations that they paid too much. As a result, vaccine shipments and vaccinations in the EU lagged those in the US; by the first quarter of 2021, the American economy was growing at an annualized rate of 6.4 percent,[11] while the EU economy was still contracting.[12]

Offer Something Money Can't Buy: Data on Real-World Effectiveness

Israel's data-rich healthcare system also played a role in convincing Pfizer to prioritize Israel in its allocation of shipments. The Israeli universal healthcare system consists of four large, nonprofit, competitive healthcare "funds" that act as health maintenance organizations (HMOs) in providing both insurance and healthcare delivery. Each of these organizations operates a network of clinics throughout the country, providing primary and specialist healthcare. The funds are supported by their own hospitals as well as government hospitals and independent ones. By law, each Israeli citizen must register with one of these funds, and each citizen's unique national healthcare ID number feeds data into an advanced digital healthcare management and electronic record management system that interconnects the entire country.

Such robust data availability gave Israel a key bargaining chip with vaccine makers: It could offer crucial data on vaccine efficacy, safety, and the benefits of herd immunity.[13] In being a lead customer that could quickly vaccinate the majority of its population, Israel could become an implicit case study that showed other nations the powerful advantages of nationwide vaccination in getting both life and the economy back to normal. Naturally, this would translate into more revenue for the vaccine makers.[14]

Three Keys to an Effective Rollout

Gil Epstein, professor of economics and dean of social science at Bar-Ilan University outside Tel Aviv, described how Israel's successful rollout of the vaccine stemmed from "obtaining a large number of vaccines and efficient distribution."[15] Acquiring the millions of doses needed to vaccinate all Israelis who wanted the vaccine was just half of the story. The other half was in planning the delivery, training volunteers for massive phone banks, developing systems, enlisting military medics, preparing and qualifying vaccination centers, and so forth.

For the delivery side of the rollout, Epstein listed three advantages Israel enjoyed: "Number one is we have Teva, which is a big pharmaceutical company that sits in the middle of Israel."[16] Teva's distribution center offered an ideal location for vaccine logistics in being (a) a few miles from Israel's biggest international airport where vaccines arrived, (b) on the outskirts of Israel's largest urban area (containing nearly half the country's population), and (c) next to a major highway, making it possible to reach anywhere in the country in a few hours. The Teva facility held 30 large ultra-cold freezers capable of storing 5 million doses. On December 9, 2020, the first trial-run shipment arrived to test the handling, transportation, and storage procedures as a prelude to much larger volumes in the following days. "We are talking about a huge estimated scope of 4 million vaccine [doses] by year's end," said Yossi Ofek, general manager of Teva Israel, to Israel's Army Radio.[17]

"Number two is they started with the medical service," Epstein continued.[18] Vaccinating healthcare workers made both pandemic-prevention sense and practical sense. In prevention terms, healthcare workers faced high risk of exposure to Covid, and a Covid-infected healthcare worker posed a threat of exposing many others to the virus. In addition, reducing the number of Covid cases among healthcare workers would reduce the chance of a collapse of the healthcare system. In practical terms, the healthcare population was an easy trial of the vaccination systems because healthcare facilities were able to quickly vaccinate all their workers.

Israel largely followed the priority system advocated by the WHO: (1) healthcare workers, (2) the elderly, (3) adults with comorbidities (4) everyone else.[19] Given its unique security challenges, Israel also vaccinated its troops relatively early. In early March 2021, Major General Itzik Turgeman, the head of the military's Technology and Logistics Directorate, told reporters, "After 10 weeks, I can declare that the IDF is the first military in the world to reach herd immunity."[20]

Vaccinating those most susceptible to severe Covid ensured the quickest reductions in hospitalizations and deaths. By rapidly decreasing the hospitalization rate through vaccinating the vulnerable, the system freed up healthcare workers for the vaccination campaign. Israel's priority system also benefited from its simplicity, designed to inoculate the highest number of people in the shortest amount of time. The system provided an ample number of sizable vaccination sites and intentionally set the "priority bands" of those eligible for vaccination to be very broad. Thus, each band included a large number of people, ensuring a steady supply of customers at the mass vaccination sites.

"Number three," said Epstein, "is all the HMOs are very, very efficient."[21] As of 2020, Israel had the third-most efficient healthcare system in the world (behind Hong Kong and Singapore) on the Bloomberg Health-Efficiency Index.[22]

Israel also benefited from volunteers such as Esti Marian. The 67-year-old Tel Aviv resident had been volunteering in an HMO-funded clinic for more than six years, fulfilling administrative and clerking roles. In December 2020, she joined a legion of other volunteers to staff phone banks supporting the mass vaccination campaign. Although Israel set up a national appointment website, the site crashed within the first 24 hours. The phone banks that were originally intended for people who simply preferred telephoning their trusted HMO also handled the surge of people who could not get an appointment on the crashed online system.

When HMO members called their local clinic and gave their healthcare ID number, people like Esti could immediately see the caller's general details on their electronic health record (EHR), such as age, address, family situation, employment, and so forth. (However, she could not see any medical details, which are visible only to medical professionals.) Following instructions from the Ministry of Health, she made appointments for age-eligible callers while directing younger ones with comorbidities to their primary care physicians to obtain a special approval for early vaccination.[23] Once the appointment website was back up and running, people could make appointments themselves. The system would then show available appointment times and locations, balancing the load among vaccination sites.

Once the person arrived at the vaccination site, it took only three to five minutes to get the vaccine into their arm. Again, the use of the electronic ID number eliminated any need for time-consuming paperwork or payment (the vaccine was free to every citizen), and vaccine site processes moved quickly. The authorities understood that vaccination is a numbers game, and the goal was to get as many people as possible vaccinated promptly.

The Pfizer–BioNTech frozen vaccine has a five-day refrigerated shelf life after it's thawed, and only a six-hour life after being diluted for injection.[24] If anyone were to miss their appointment, the dose allocated to them would be left over at the end of the day and potentially discarded. Similarly, if a clinic were to overestimate its five-day demand level, it too may have leftover doses. To avoid wasting these doses, Israel's vaccination centers put out text-message blasts toward the end of each day inviting everyone—regardless of age or medical condition— to come get vaccinated. Those who were eager to be vaccinated learned to linger near clinics that might offer an end-of-day shot.[25]

From Rolling Out to Opening Up

Less than a week after starting the vaccination process, Israel ramped up to vaccinating 1 percent of its population each day. By contrast, the US took over four months to reach this rate, which was as high as the US's vaccination process ever got.[26] At the peak of Israel's vaccination campaign, a month after it started, Israel was vaccinating another 2.1 percent of its population every day.

At the height of the pandemic in mid-January 2021, Israel was suffering from more than 900 new cases and 10 deaths per day per million citizens. As total vaccinations grew, case rates subsided. By the end of May 2021, the country had reached nearly 60 percent fully vaccinated citizens, with 80 percent of people over 60 fully vaccinated. Israel reopened much of its economy, and daily new cases had fallen more than 99 percent to three to five per million,[27] with nearly zero deaths.*

Lessons Learned the Hard Way

Many countries bungled what was an elementary math problem in basic operations management: estimating and acquiring the resources (doses, syringes, shipments, personnel, vaccination stations, etc.) needed to vaccinate their populations in a reasonable timeframe. For example, India's purchase of 350 million doses ranks high in absolute numbers, but relative to its 1.4 billion citizens, it's a scant coverage at 12.5 percent.[28] As Covid surged in India in the spring of 2021, many criticized India's large domestic vaccine makers, such as the Serum Institute of India (SII) and Bharat Biotech, for insufficient capacity. Adar Poonawalla, CEO of the

---

* In late summer 2021, Israel experienced another wave of Covid cases, with the unvaccinated elderly more than five times more likely to experience a severe case of Covid-19 than their immunized counterparts.

SII, told the *Financial Times*, "I've been victimized very unfairly and wrongly," explaining that he had not boosted capacity because "there were no orders, we did not think we needed to make more than 1 billion doses a year."[29] Unfortunately, India's lack of vaccines coincided with a deadly pandemic spring wave that killed untold numbers of Indians. Weak healthcare systems, insufficient oxygen supplies, and a lack of political leadership all led to horrendous suffering and death on a staggering scale compounded by the lack of vaccines.

In contrast, the US spent billions up front funding the development of vaccines and ensuring priority access to supplies for more than enough vaccine to cover its citizens. But then, it launched one of the biggest failures of logistics planning and execution in the modern era. According to epidemiologist Michael Osterholm, director of the Center for Infectious Disease Research and Policy at the University of Minnesota, "What happened is, and I give a great deal of credit to Operation Warp Speed, especially with the R&D and getting it manufactured, but they didn't account for the importance of the last mile, which is distribution."[30] Operation Warp Speed had an original goal of 20 million vaccinations by the end of 2020. On bulk deliveries, it missed this goal by 30 percent (only 14 million were delivered). But the actual vaccination rate was missed by more than 85 percent (only 3 million people were vaccinated by the year's end).

Whereas Israel has a strong central government to oversee and control the end-to-end procurement, distribution, and administration of the vaccines, other countries were victims of their decentralized and fractious political structures. The US, for example, was founded as an amalgamation of 13 independent colonies bounded by a constitution delegating all governance to these 13 states in matters not explicitly designated within federal purview. Although the federal government could ensure the development and production of large volumes of vaccines, each of the 50 states governed its own rollout, and in some states, that governance was further fragmented by delegation to county or municipal health departments. The result was scattershot efforts in which tens of millions of doses sat unused during the early weeks of the US vaccination campaign.[31]

Mismatched Planning for Supply and Demand

Other failures came from a lack of sales and operations planning (S&OP) processes that consumer product organizations have long used to coordinate both supply activities to match expected demand and sales activities to match available supply. Lack of this kind of coordination meant that some states in the US opened eligibility too fast given the overall limited supplies of vaccines coming from the manufacturers. The resulting surge in demand for vaccines caused several problems that hampered the rollout.[32] First, the most vulnerable citizens (e.g., the elderly) lost out to younger, determined, tech-savvy seekers of appointments and individuals who gamed the system by misrepresenting their medical histories.[33] Second, overwhelmed sign-up systems, the difficulty of getting appointments, and long queues made for mass frustration, reinforced by unflattering news reports—all resulting in a reduction in the number of people willing to seek the vaccine. "In the rush to satisfy everyone," said Dr. Rebecca Wurtz, an infectious disease physician and health data specialist at the University of Minnesota's School of Public Health, "governors satisfied few and frustrated many."[34]

Worse, the lack of coordination between local vaccination sites, state vaccine allocation managers, and the federal vaccine distribution system wreaked havoc at these vaccination sites. For example, Dr. Patricia Schnabel Ruppert, health commissioner for Rockland County, New York, just outside New York City, explained, "Right now, I can't book vaccine appointments for next week or the following week. We don't know how many doses we'll get, so we don't know how to organize staffing or how many volunteers we need."[35]

Stretching the Limited Supplies

In an effort to partially protect as many people as possible, a number of countries chose to give more people a single dose of vaccine immediately rather than fully immunize fewer people with the requisite regimen of two doses over the vaccine maker's recommended three- or four-week period. These included Canada,[36] the UK,[37] and several other European countries.[38] These countries planned to delay the second dose up to 12 weeks, when vaccines would be more readily available.

This strategy was quite controversial in scientific circles,[39] and other countries, such as the US, rejected it. At issue was the gulf between speculative plausibility and objective science regarding whether a single dose offered adequate levels of partial protection and whether delaying the second dose would hinder developing fuller immunity. A British study, using real-world data (in contrast with vaccine trial data or extrapolated data) estimated that a single dose of Pfizer–BioNTech vaccine offered only 33 percent protection against symptomatic disease from the Delta variant, in stark contrast to the 88 percent effectiveness of two doses (two weeks after the second dose).[40] Other studies documented higher single-shot effectiveness. For example, the FDA released a study documenting significant protection after one shot: 52.4 percent for the Pfizer-BioNTech vaccine and 69.5 percent for Moderna.[41] (Note that simple math suggests that if a single dose is only 33 percent effective, then single-dosing twice as many people is worse than a double-dose for half as many. On the other hand, if the single dose is 52 percent effective, then single doses actually are better.)

Another strategy to stretch limited supplies and reduce the complexity of coordinating shots was to mix and match vaccines. If not enough of a particular vaccine was available, one's second shot could come from a different manufacturer. Canada, Italy, South Korea, and a few other countries advocated this approach. Leading politicians such as Angela Merkel, Mario Draghi, and Justin Trudeau personally chose different second vaccines for their respective second shots to demonstrate the safety and efficacy of the approach and to show that both vaccines where equally good in the eyes of the countries' leaders. Some research suggested that this strategy could even produce higher immunity levels than the standard two doses of one vaccine. However, questions remained.[42]

Disappointments in Finding Appointments

US fragmentation extended to the bewildering array of options for vaccination sites offered by states, counties, cities, hospital chains, clinics, pharmacies, and grocery stores. Unfortunately, in many states, the only way to get appointments at any of the many different provider locations was to visit the websites of each provider, create an account, and continually check back, hoping to snag one of the scarce appointment slots. That fragmented delivery system forced consumers to spend hours scrambling to book appointments for their elderly parents, relatives, and then themselves. The fragmented system also created demand volatility and planning problems. Eager consumers signed up on multiple websites, creating an illusion of demand. In some cases, consumers would grab the first appointment they could (which might be quite far away) and then try to find better, closer appointments. The result was a large number of no-shows.

A few states, however, bucked the US penchant for decentralized vaccination management with significant success. West Virginia, for example, managed the process centrally. All vaccines that arrived in West Virginia were shipped to five hospitals in various parts of the state. These hospitals served as distribution hubs, getting vaccines to local health centers, clinics, doctors' offices, and other hospitals, while the state tracked the operation through a central registration system. By the beginning of February, 11 percent of West Virginians had received their first shots, compared to 7.2 percent in Massachusetts and 7.7 percent in California—both states with much better-funded healthcare systems.[43†]

Taming the Chaos

Despite Israel's centralized healthcare system and months of planning,[44] the first few days of its vaccine rollout were also chaotic.[45] Early on, both the call centers and digital tools were overwhelmed. Appointment seekers were given conflicting information about which local vaccination centers would be open and when. People, including the elderly, were being told to travel long distances to reach an available appointment. Some vaccination centers—especially in large cities—had long queues due to overbooking, and others had doses left over at the end of the day due to less-than-expected demand. Within days, however, the website was fixed, communications were made consistent, and vaccination centers started announcing surplus vaccines available to any citizen at the end of each day.

Dr. Hagai Levine, an epidemiologist at Hebrew University–Hadassah School of Public Health in Jerusalem, told Vox in January 2021 how Israel managed: "For a vaccination campaign, we are well-prepared, but we're also flexible. When you plan, you don't know, for example, how the cold chain will look, how many vaccines you will get—so you need to make rapid adjustments. And we are good at that."[46] In a high-speed, large-scale rollout of a new product, planning provides some assurance that nearly the right amounts of nearly the right resources

---

† Unfortunately, probably due to relatively low vaccination rates and no mask mandate, by September 2021, West Virginia had the highest rate of Covid-19 cases per capita among all US states.

will probably be near the right place. But then, the actual patterns of supply, demand, and operational effectiveness will dictate rapid adjustments to the plan. As heavyweight boxer Mike Tyson once said when asked about his opponent's plan for the upcoming fight, "Everybody has a plan until they get punched in the mouth."[47] More to the point, US President Dwight Eisenhower said, "Plans are worthless, but planning is everything."[48] What he was saying is that any plan for a crisis or an emergency situation is likely to be incorrect, but the planning process demands the thorough exploration of options and contingencies. The knowledge gained during this probing is crucial to the selection of appropriate actions as future events unfold.

Mass Confusion

In my own state of Massachusetts, the initial chaos lasted for weeks—and then months—owing to flaws in system design, distribution strategy, and allocation systems. In mid-February 2021, the state's vaccination appointment website and telephone hotlines suffered an absolute, colossal failure under the load of a million newly eligible citizens.[49] Experts were quick to point out that although the Covid pandemic might be something entirely new, designing websites to handle surges of millions of users was not. "These are known problems that you can design around," said Jay Jamison, an executive at Quickbase, a Boston-based web technology company.[50]

The main problem with Massachusetts's attempts to develop a vaccination website was that, at first, it didn't think it needed one. The original government's plan was based on local sign-up rather than a centrally managed process. In doing so, Massachusetts abandoned the emergency mass vaccination plan it had had in place for 20 years and hired consultants. Said one local public health official, "They took the playbook, threw it in the dumpster, and privatized the whole thing."[51]

The web designers then borrowed the web/user experience strategy used successfully for flu vaccination campaigns without understanding the differences between flu and Covid. In the case of the seasonal flu, the state had always had enough vaccine, and citizens felt less of a pressing need to sign up for a flu shot as soon as possible. At any given time in the run-up to flu season, only a modest number of citizens would be on the site trying to make an appointment, and most of them readily found an appointment on the first try.

In contrast, with Covid, the shortage of vaccines, the priority group system, and dramatically greater urgency combined to crash the appointment system. As each group became eligible, they flocked to the website in large numbers, quickly exhausted the available appointments, and then hammered the site by repeatedly trying and failing to snag an appointment. As any retailer knows, reported or perceived shortages of a key item (e.g., toilet paper in the spring of 2020) has the unfortunate effect of increasing short-term demand and making the shortage panic worse.

Under such conditions of a severe supply-demand mismatch, a better process would have been a pre-registration and invitation system that steadily meters the available appointments by priority and lottery. (First-come-first-served only fosters a frantic panic.) Thus, people could pre-register anytime on the site with their information and preferences, the system would sort people by priority, use a lottery for fairness among people in the same priority group, and then text, email, or call people when their turn came and a nearby site had vaccines. The numbers of sent invitations could match the numbers of newly available appointments, ensuring a high chance that people could find an appointment easily.

Massachusetts also had a complicated and confusing priority system that included four phases with many subcategories in each phase.[52] The first phase had six subphases including (in order): healthcare workers directly serving Covid patients, workers in long-term care facilities, first responders, workers in congregate care settings, home-based healthcare workers, and, finally, non-Covid-facing healthcare workers. The second subphase started with people over 75, then people over 65 as well as residents of low-income housing, and a slew of other categories including people with any two of some 18 specified comorbidities. This was followed by K–12 educators, followed by residents 60 years or older who work in one of 12 specified sectors, followed by residents 55 and older with a single specified comorbidity. And these were just the priorities of Phase 1.

Not only did the complex system create confusion about who was eligible at any given time, but it also created frustration and the specter of line jumpers. Neither the state nor the vaccination sites were able to verify the status of most people claiming to be in categories related to employment, comorbid conditions, and the like. Furthermore, some categories, like being a companion for an elderly person (75+) spawned a black market in which vaccine-seeking younger people offered hundreds of dollars to older people to escort them to the vaccination site whether they needed help or not, so they too could get vaccinated.[53]

Right from the beginning, Massachusetts was particularly concerned about equitable access to vaccination and therefore distributed the vaccine widely throughout the state to large numbers of smaller outlets. And as any logistics professional knows, smaller outlets suffer from greater demand uncertainty compared to large outlets.‡ Small outlets have more problems with unpredictable surpluses (which means excess vaccines may have to be discarded) and shortages (which means that people will have to be turned away). Indeed, after the initial period, Massachusetts pivoted to focus on large mass vaccination centers such as in the Boston Red Sox's Fenway Park, the New England Patriots' Gillette stadium, and several convention centers.[54] Large-scale sites help average out fluctuations in both demand and capacity.

Over time, Massachusetts also improved its website and distribution by adding a system for preregistration and notifications. This system, however, only applied to the state's mass vaccination sites. Appointments at pharmacies and retail locations remained as fragmented as before.[55] Consequently, people were still going to the websites of CVS, Walgreens, Walmart, and the like to try and secure a vaccination appointment. By the beginning of May, vaccine supplies were catching up with demand, and by mid-May 2021, Massachusetts was one of the first three states to reach a goal set by President Biden of having 70 percent of adults vaccinated with at least one dose by July 4. The other two were the neighboring states of New Hampshire and Vermont.[56]

---

‡ The technical explanation is that demand volatility is a function of the coefficient of variation (CoV) of the demand. Higher CoV means higher volatility. The CoV is the ratio of the standard deviation to the mean of the demand distribution. With smaller locations, the mean is relatively small compared to the standard deviation, which makes the CoV high. Thus, locations with smaller average demand will suffer more uncertainty than large ones.

Equality of Access

Decisions about where to locate and operate vaccination sites were modulated by social priorities of who most needed access to vaccines given the spectrum of societal conditions that exacerbated both Covid-related health problems and vaccine-adoption problems. "We have a socially, politically, and racially fragmented society that is under the stress of a disease that causes illness and death," said Monica Schoch-Spana, a medical anthropologist at the Johns Hopkins Center for Health Security in November 2020, while Phase 3 trials were ongoing for several vaccine candidates.[57] "If there was ever a time when public perception of fairness and justness was important, that would be now," she added.

Later data showed that racial minorities such as Blacks and Hispanics had lower rates of vaccination despite those groups suffering higher relative rates of Covid infection and death. As of May 3, 2021, data from 42 US states showed that 39 percent of Whites had gotten at least one dose, compared to only 27 percent of Hispanics and 25 percent of Blacks.[58]

Obstacles to Vaccination

Part of the inequality of vaccination rates stemmed from fears and other obstacles to vaccination (as distinguished from the problem of false information discussed in the next chapter). The Black population's distrust of government healthcare systems has been but one impediment to achieving high vaccination rates. Other communities, such as the Amish in Ohio, Indiana, and Pennsylvania, shun modernity and believe in natural remedies. West Virginia University sociologist Rachel Stein explained, "We, as non-Amish, are more on board with preventative medicine. They certainly don't have that mindset that we need to do things to stop this from happening."[59] In addition, many undocumented immigrants may have feared deportation or (wrongly) believed they were ineligible for vaccination.[60]

Other low-income citizens faced practical obstacles related to time, money, transportation, and other obligations. About 13 million American workers hold multiple jobs,[61] leaving them with little time and even less energy to deal with the hassle of getting vaccinated. In addition, 11 million single-parent households face the added obstacle and costs of finding childcare while getting vaccinated. And about 10 million households lack access to a private vehicle, adding transportation challenges to the mix.

Twenty-five US states implemented strategies to improve vaccination access and equity[62]: for example, adjusting vaccination center locations, vaccine allocation, eligibility standards, and appointment slot availability to favor underserved groups or neighborhoods. Many states created call centers or text-message systems to reach those without internet access. Wider outreach and communication strategies publicized that the vaccine was available, safe, and free. Many of these efforts collaborated with community groups and local leaders to deliver culturally competent and linguistically appropriate information through trusted messengers.

Eventually, the US managed to create a broad downstream distribution network leveraging retail outlets found in almost every community. The Federal Retail Pharmacy Program for Covid-19 Vaccination, enlisted pharmacy and grocery store chains, adding more than 38,000 convenient vaccination locations to the usual array of healthcare clinics, hospitals, community health centers, and mass vaccination centers.[63] For example, CVS alone had nearly 10,000 outlets, as well as nearly 100,000 pharmacy technicians and pharmacists who could deliver injections. CVS also said that 70 percent of the US population lived within three miles of one of its outlets.[64] As of late September 2021, this federal program had delivered nearly 133 million shots, of which 43 percent had gone to racial or ethnic minorities.[65]

The Double-Edged Sword of Internet Access

In theory, the internet should have substantially improved access to vaccination. The internet and smartphones made it easier for people to find vaccination sites, determine vaccine availability, make appointments, receive notifications about available appointments, and use GPS to navigate to appointments in places they had never visited before. However, in practice, the internet advantaged wealthier, educated people.

The same technology skills that enabled high-income workers to avoid job losses during the pandemic also enabled them to efficiently navigate all the complex online systems for securing appointments. Websites with first-come-first-served allocation for scarce appointment slots naturally favored people with faster computers, faster internet connections, and better technology skills. Tech-savvy vaccine seekers used automated bots created by clever programmers to help repeatedly check websites far and wide for appointments.

In contrast, many low-income and minority populations, who were disproportionately impacted by Covid both economically and in terms of health, largely lacked the computer equipment, broadband access, and personal technology skills to compete for coveted, hard-to-get online appointments.[66] Many appointment sites required email accounts and cell phone numbers (for text-message alerts) that many elderly and poor simply do not have.[67] When authorities opened up call centers to help those without computers, tech-savvy vaccine seekers soon learned of the call centers and exploited them too. Even when governments intentionally opened vaccination centers in neighborhoods with under-served populations, vaccine enthusiasts from wealthy neighborhoods learned about them from Twitter or Facebook and promptly snatched up all the appointments.

Some vaccine administrators tried to combat the problem of tech-savvy non-locals monopolizing the vaccines. For example, when a vac-cine clinic in rural western Massachusetts discovered that 95 percent of its appointments were for outsiders who had snatched up all the appointments in minutes, authorities canceled the appointments and created a more private system for locals.[68] Short of requiring an ID to prove one was from an underserved geographic area (a counterproduc-tive policy likely to scare off those who distrust the authorities), it is difficult to keep a vaccination center's existence a secret from those who are determined to find it.

Inequality on a Global Scale

Compared to the vaccination inequalities within the US, the situation was far more pronounced on the international stage. Rich countries all but monopolized the earliest batches of the vaccines. As of mid-May 2021, whereas the US had delivered more than 80 doses per 100 citi-zens, South America had delivered only 22, India had delivered 13, and Africa less than two doses per 100 citizens.[69] On the procurement front, the EU lined up contracts for nearly seven doses per citizen and the US made deals for nearly four doses per citizen, while the African Union had secured a scant 0.36 doses per citizen. Overall, by the middle of May 2021, high-income countries had secured about two-thirds of all contracts for vaccines.[70] To illustrate this, a coalition of NGOs called The People's Vaccine Alliance claimed that, in March 2021, rich nations were vaccinating at a rate of one person every second while most of the poorest nations had yet to receive any vaccine at all.[71]

The Negatives of Nationalism

As countries ramped up the vaccination of their citizens, a new cause of supply constraints arose. International clashes over vaccine access and supplies led countries to place trade restrictions on vaccine exports. For example, India suspended vaccine exports in March 2021 (even before the April–June explosion of the pandemic there) in order to preserve doses for its own population.[72] At the same time, the UK and EU were embroiled in a dispute over the allocation of vaccines made by AstraZeneca (a British-Swedish company with many facilities in the EU).[73] The Covid-19 pandemic had previously sparked other trade restrictions: More than 70 jurisdictions instituted a combined 137 export curbs on personal protective equipment (PPE) and other medical products in 2020, according to watchdog Global Trade Alert.[74]

The Fences Created by Defenses

During the pandemic, the US government used the Korean War-era Defense Production Act (DPA)[75] to prioritize, mandate, and fund production of key items. In the early stages, the US used the DPA to accelerate the production of PPE and ventilators.[76] Later, the government used the DPA to aid vaccine makers in acquiring bioreactor bags, filters, pumps, vials, and raw materials.[77]

The DPA does not explicitly block exports; instead, one of its legal tools is the use of so-called rated orders to mandate that manufacturers and suppliers prioritize certain orders above all others—i.e., it preempts the company's normal allocation rules. However, if total demand exceeds total supply, then rated orders all but prohibit exports or even fulfillment of orders to other domestic companies or for other products.

For example, DPA-rated orders for other vaccine makers meant that even Pfizer faced delays in getting key materials and ingredients for vaccine production.[78] Pfizer then sought US government help in getting higher-priority DPA access. It finally traded more vaccine supplies to the US in return for access to several specialized vaccine manufacturing supplies.[79] Another consequence was that the DPA created shortages of some other drugs, including antibiotics and drugs used for thyroid conditions and breast cancer. "It's very much a blunt instrument to be used, when you may need a scalpel," said Kelly Goldberg, a vice president in the law division of Pharmaceutical Research and Manufacturers of America, a pharma industry trade group.[80]

The problem is that the world's vaccine producers depend on each other for ingredients, since they operate complex multinational supply networks of chemicals, fatty acids, glass vials, and a host of other components. As the Trump administration found out when trying to restrict the trade of medical supplies in response to the pandemic in 2020, countries do retaliate. The result is that everybody's supplies become restricted and worldwide vaccine production falls. There are 11 vaccine-producing countries in the world plus the EU. They source an average of 88.3 percent of their ingredients from the other vaccine-producing countries and thus mutually depend on each other for supplies. (The biggest exporter of ingredients is the US, followed by the EU and the UK.) Any trade dispute can reduce and even halt vaccine production.[81]

Heavy-handed restrictions also directly harmed the restrictive countries. Nathalie Moll, the director-general of the European Federation of Pharmaceutical Industries and Associations (EFPIA), told Portuguese weekly *Expresso,* "We are losing time already; we don't see a need for this at all. It just slows us down at the moment."[82] Similarly, the EFPIA said in a statement, "Risking retaliatory measures from other regions at this crucial moment in the fight against Covid-19 is not in anyone's best interest."[83]

Vaccine Diplomacy

While some countries hoarded vaccines and blocked their export, other countries used vaccines as strategic currency for diplomatic and geopolitical ends. Countries such as China, Russia, and India have their own mature biotech industries that developed and mass-produced a number of Covid vaccines. Those countries' governments used these vaccines to both immunize their own people and to offer doses to other selected countries to create or strengthen diplomatic relationships.

As of May 2021, four Chinese companies had created five China-approved vaccines, including two that gained approval in dozens of countries around the world.[84] The Economist Intelligence Unit estimated that, by the end of April 2021, China was sending its vaccines to 90 countries.[85] China has been using its generosity in sharing vaccines to bolster its diplomatic aims. These include cultivating allies who will oppose other countries' criticism and interference in China's internal affairs, allowing China to mine natural resources on their land,[86] and acquiescing to China's expansion into the South China Sea.[87]

Russia named its country's vaccine Sputnik V, referring to the Soviet Union's 1957 launch of the world's first satellite.[88] The vaccine was developed by Gamaleya Research Institute of Epidemiology and Microbiology in Moscow, an institute within the Russian Ministry of Health. As of early August 2021, Sputnik V was approved in 70 countries.[89] Russia contracted with more than a dozen manufacturers in 10 countries (including India and China) with plans to produce a total of 1.4 billion doses.[90] According to an Economist Intelligence Unit estimate in April 2021, Russia planned to send vaccines to 70 countries, many of which were also getting Chinese vaccines.[91] In addition to geopolitical objectives, Russia was seeking access to natural resources, such as lithium in Bolivia, and additional development projects.[92]

India has its own homegrown vaccine (Bharat Biotech's Covaxin) as well as licenses to make several of the leading vaccines (Oxford–AstraZeneca, Johnson & Johnson, Novavax, and Sputnik V).[93] For example, the Oxford–AstraZeneca vaccine is manufactured by the Serum Institute of India, the world's largest maker of vaccines, and sold under the brand name Covishield.[94] India is also part of a four-country alliance known as the Quad (the US, Japan, Australia, and India) that is working to finance, produce, and distribute a billion doses to Asian countries.[95] India's Vaccine Friendship program "raised India's standing and generated great international goodwill," according to India's Foreign Minister S. Jaishankar,[96] by providing gifts of vaccines to its neighbors,[97] island nations, African nations, and allies of Taiwan.[98]

Despite India's good intentions, by the end of spring 2021, its efforts to help vaccinate the world had stalled; India was suffering from a tidal wave of infection and the government blocked vaccine exports, forcing its vaccine makers to redirect all their production for domestic use.[99] It became clear that, among its other failures, the Indian government had not ordered enough vaccine for its 1.4 billion citizens.

Diplomacy's Difficulties

Vaccine diplomacy is not without its problems for both donor and recipient. Brazil's regulator, ANVISA, refused to approve Sputnik V, saying the data showed the vaccine carried an unacceptable risk of having live, replicating adenovirus contaminating the batches of the non-replicating viral vector.[100] Some of the problem could have come from a misunderstanding of the technical documentation as well as quality control process differences across the different vaccine manufacturing sites, but it was exacerbated by a lack of transparency. Rather than address the concern and provide additional requested data, the Russian backers of the vaccine decided to sue for defamation.[101] An EU report accused that Russia, as part of its efforts to aggressively promote Sputnik V, was spreading misinformation that the West was trying to undermine its vaccine.[102]

Many have been skeptical of the Sputnik V vaccine given that (1) Russia's President Vladimir Putin approved its use in August 2020 after testing it in only 76 people—and before Phase 3 trials had even begun,[103] (2) Russia touted good results based on insufficient samples sizes,[104] and (3) Russia has not been transparent about its vaccine development, clinical trials, and production systems.[105] Sputnik V may well be a safe and effective vaccine, but its use for ulterior geopolitical purposes tainted the efforts of Russian science to contribute to the global vaccination effort.

When Russia convinced Slovak Prime Minister Igor Matovič to buy 2 million doses of Sputnik V, the deal seemed like a political win for both sides; Slovakia hoped to halt its high per-capita death toll and Russia desired a beachhead into a Europe that had not approved Sputnik V. However, the Prime Minister's governing coalition partners reacted badly to the secret deal, which ran counter to the country's pro-EU and pro-NATO stance. The political turmoil forced Matovič to step down, which, in sowing dissent in the EU, may also have been a strategic win for Russia. Slovak foreign minister Ivan Korčok described Russia's efforts as a "tool of hybrid war" that "casts doubt on work with the European Union."[106]

In other cases, vaccines that showed early promising signs produced less than stellar results. Brazilian trials in January 2021 of the Chinese biotech company Sinovac's CoronaVac measured its effectiveness at only 50 percent,[107] which may be acceptable as a better-than-nothing option but is clearly inferior to several other vaccines boasting effectiveness of 90 percent and above. Some 20 countries had collectively ordered nearly half a billion doses of CoronaVac by the end of May 2021, including 125 million for Indonesia and 100 million each for Brazil and Turkey.[108] Yanzhong Huang, a senior fellow at the Council on Foreign Relations and an expert on healthcare in China, told the *New York Times*, "Those countries that have ordered the Chinese-made vaccines are probably going to question the usefulness of these vaccines."[109] The United Arab Emirates and Bahrain were going so far as to offer a Pfizer–BioNTech booster shot[110] to those who got the Chinese Sinopharm vaccine over concerns about efficacy, especially as variants of the original virus spread globally.[111]

From Hoarding to Handouts

Vaccine diplomacy is likely to expand as richer countries reach the saturation point in their vaccination campaigns and start using their surplus doses for diplomatic ends, too. Dr. Anthony Fauci, the US government's top infectious disease expert, said at a White House news conference at the end of March 2021, "After we do take care of the really difficult situation we've had in our own country with over 535,000 deaths,§ we will obviously, in the future, have surplus vaccine, and there certainly is a consideration for making that vaccine available to countries that need it."[112]

As the spring of 2021 turned to summer, rates of new vaccinations in the US tailed off, vaccine supplies in the US began to exceed demand, and the US government began switching from hoarding to handing out doses to other countries. By May, the US had announced plans to send 80 million doses of its surplus vaccines to other countries by the end of June 2021,[113] in addition to 4 million doses already sent to close trading partners Mexico and Canada that March.[114] (These first donations included 64 million AstraZeneca doses the US had bought, but which were not approved by the FDA). In June, the US announced an even larger donation of 500 million Pfizer–BioNTech doses to be distributed by COVAX to lower-income countries that were struggling to vaccinate their people.[115] In September, President Biden announced an increase of the US vaccine donation to 1.1 billion vaccines.[116] This contribution, which was significantly larger than all other countries combined, was used by the Biden administration to justify its decision to provide booster shots, despite some internal opposition. Clearly, politicians and

---

§ The US death toll has risen substantially since then.

nations are using the crisis to further both domestic and international agendas.

4. Encouraging Customer Adoption

While vaccine makers, governments, and logistics companies worked hard to invent, produce, and distribute billions of vaccine doses, getting those doses into the arms of people was up to those people themselves. Continuing the product launch analogy from the last chapter, a new product's adoption by customers depends on a wide range of factors, such as the product's cost-benefit trade-offs for individuals, companies, and governments. Central to the dynamics of adoption is product information and the interplay of true, false, and debatable "facts" about the product.

The Advantages of Adoption

Vaccination offers benefits to both individuals and society as a whole. At the individual level, vaccination reduces the chance of being infected with Covid and especially reduces the possibility of severe infection and death. At a more macro level, vaccination promises an end to the scourge of Covid-19 and all the associated disruption and suffering felt by people, the economy, and normal life. However, the disease would only subside quickly if enough people got vaccinated to reach herd immunity.

What's in It for Me?

Individuals who get vaccinated benefit personally, because vaccination provides a level of immunity against catching Covid. That immunity substantially reduces their chances of getting sick, being hospitalized, potentially dying, or suffering from "long Covid,"* which seems to affect about one in three people with symptomatic disease.[1]

Vaccination isn't the only way to develop immunity—catching Covid is the competing "product" that can provide immunity from catching Covid again. However, of the two ways to gain immunity, vaccination is far more preferable by any measure of preventing symptoms, side effects, hospitalization, and death—not to mention the associated monetary costs. At the peak of its vaccination campaign, the United States was able to immunize more than 3 million people per day via vaccination, with a low incidence of side effects. In contrast, during the January 2021 pandemic surge, about one-tenth of that number were infected daily, offering natural immunization to those who recovered. However, this natural immunization also killed 3,500 people per day in that same month and swamped the nation's hospitals with the seriously ill.[2]

What's in It for My Family and Friends?

An individual also benefits from vaccination by reducing the chances of spreading Covid to their family, friends, and coworkers.

That risk of spread depends on the disease's infectiousness. Recall from Chapter 1 (p. 1) that epidemiologists measure by the infectiousness by R_0 (also known as the *basic reproduction number*). More specifically, R_0 is the average number of individuals (in a susceptible population) to whom an infected person will transmit the virus during the time that they are contagious. R_0 of a respiratory disease depends on the biological abilities of the pathogen to reproduce in the human body and escape in droplets and, particularly, aerosolized bodily fluids to expose others. It also depends on a wide range of social attributes, practices, and behaviors (like crowding, hygiene, mask use, and ventilation) that affect people's exposure to these pathogenic droplets and aerosol emissions. For example, the R_0 on a cruise ship may be higher than the R_0 in a dense city, which in turn may be higher than the R_0 in a rural area.

---

* Sufferers of "long Covid" are those who continue to feel symptoms long after the days or weeks that represent a typical course of the disease. Interestingly, it tends to affect younger patients and, in some cases, people who suffered only mild symptoms.

Typical values of vary by disease. Seasonal influenza typically has an R_0 = 1.2–1.4, while more serious pandemic strains of influenza might have R_0 = 1.3–2.0. At the other end of the contagiousness spectrum is measles, with R_0 = 12–18. The original variant of Covid in Wuhan (the ancestral strain) had R_0 = 2.4–3.4. However, as Covid spread worldwide, it adapted to people's attempts to control it. Subsequent variants were more infectious: The Alpha variant has R_0 = 4–5, and the Delta variant has R_0 = 5.0–9.5.[3]

What's in It for Us?

As more and more people gain immunity to a disease through vaccination or natural infection (and recovery), the prevailing reproduction number is no longer the basic R_0 value. Instead, epidemiologists talk about the *effective reproduction number*, R_e, which is the reproduction number once a portion of the population has gained immunity and is less susceptible. Recall that the value of the basic R_0 is defined for a fully susceptible population before any measures against the pandemic are taken. As more and more people get vaccinated, the value of R_e will decrease further and further.

Assume, for example, that R_0 = 3, as in the case of the original Wuhan variant, and in addition, one-third of the population is immune. In this case, only two out of three people, on average, will be susceptible, and only two in the average group of three people exposed to an infected individual will actually catch Covid. With this level of immunity, the effective reproduction number will be only R_e = 2, and the disease will spread more slowly than when no one was immune.

In this example (with R_0 = 3), if immunity increases further to more than two out of the three people (66.7 percent), then only one person, on average, will be infected by each infectious carrier. In this case, R_e = 1, and the number of new infections will stay constant and not increase geometrically. If still more people are immunized, R_e will decline below 1, meaning the community will reach herd immunity, the disease will eventually disappear, and life will return to normal.

If a vaccine is at least 95 percent effective, then vaccinating about 70 percent of the population ensures that about two-thirds are immune. The actual level of immunity in the population will likely be higher, because some unvaccinated individuals will have become immune by catching and recovering from the disease. However, if the original is R_0 higher, as in the case of the Alpha and Delta variants, then reaching herd immunity requires an even higher rate of immunity in the community. Similarly, if a vaccine has lower effectiveness, then the percentage of people vaccinated must increase.

"If we have a 95 percent effective vaccine and only 40 percent to 50 percent of the people in society get vaccinated, it's going to take quite a while to get to that blanket of herd immunity that's going to protect us," said Anthony Fauci at Harvard's T.H. Chan School of Public Health.[4] What was left unsaid is that if only half the US population is fully vaccinated, a large fraction of the other half will have to suffer through Covid in order for the nation to reach herd immunity, with many more infections and deaths in the process.

We Built It, But Will They Come?

A key part of any new product launch is reaching the target customer segments and convincing them to embrace the product. Customer adoption depends on the customer's intrinsic beliefs about the product, what the product delivery organization does to convince or entice customers, and information or opinions created by third parties whom customers listen to and trust. In terms of intrinsic beliefs, Covid-19 vaccines saw an extreme spectrum of people's levels of interest—from wealthy vaccine enthusiasts willing to pay £25,000 for vaccine tourism in Dubai[5] to the militant anti-vaxxers who were—and remain—convinced that the vaccine contains Bill Gates's 5G-controlled microchips and mind-control chemicals, causes autism and infertility, alters the recipient's DNA, and who knows what else?[6] In between those extremes was a range of those wanting the vaccine but willing to wait and those more hesitant to get it.

Providers of any product or service typically have five main tools for increasing the rate of customer adoption: information, influence, collaboration, convenience, and incentives. This holds true for vaccination drives as well.

Information: Show Me the Data

The first tool for increasing the rate of consumer vaccination is information, which forms the foundation of many product-adoption decisions: what the product is, how it works, what benefits it can provide, what risks it may entail, how much it costs, where to get it, when to get it, and so on. The persistently high rate of cases and deaths provided natural motivations for many to get vaccinated. For example, as Covid deaths in the US surpassed the number of US deaths in World War I, World War II, and the Vietnam War combined, more and more Americans personally knew someone who had been sickened by or had died of the disease, prompting them to seek vaccination.

Both federal and local public health authorities, as well as most local doctors stressed the safety and efficacy of the vaccine. At the same time, governments at all levels, as well as vaccination providers and media outlets, publicized the specific local plans for eligibility, vaccination site locations, special vaccination drives, and how to book an appointment. Vaccination was also billed as a means to get back to normal—not having to wear masks, socially distance, or curtail normal activities. However, information itself is not enough. As with any other messaging, an essential criterion for acceptance is that the receiver trusts the source of the information, an issue which may vary from group to group.

Influence: Whom Do I Trust?

Some people hold beliefs that are antithetical to vaccination, and they tend to disbelieve mainstream information sources. For example, in Israel, ultra-Orthodox Jews—the Haredim—tend to reject the modern world in order to focus on a life of religious study. To boost the low vaccination rates in this population, the Israeli government worked to convince religious leaders of the ultra-Orthodox sects about the merits of vaccination.[7] Haredim leaders issued proclamations supporting vaccination on the basis of being able to return to social activities (e.g., the large weddings and funerals favored by the Haredim) that had been forbidden by Covid regulations. These efforts increased the vaccine take-up rate in this vaccine-hesitant population to 78 percent.[8]

To increase the vaccination rate among the Black population in the US, former Centers for Disease Control and Prevention (CDC) director Tom Frieden argued that personal appeals from people who are most like the people who are the target population is imperative for a successful public health campaign. "The message resonates better when disseminated by familiar and consistent voices," he said.[9] In these communities, those voices took the form of community and faith leaders especially.

Collaboration: Enlisting the Local Doctor

Experiences with other public health issues illustrate the importance of local collaboration in securing adoption of new medical treatments, including vaccines. For example, the advent of antiretroviral medications to combat the spread of HIV/AIDS in Africa created its own public health challenge: how to get these new treatments to the right people in rural African areas. Most rural Africans received medical treatment from local traditional healers who saw Western medicine as a competitive threat that would take patients away from them.

"It took a while for the Westerners, investigators, and people with the antiretroviral drugs that were working in the United States and Europe to sit down with the traditional healers and create a way for everybody to win," said Dr. Howard M. Heller, Infectious Disease Consultant at Harvard Medical School and Massachusetts General Hospital, and Senior Advisor for Clinical Partnerships at the Massachusetts Consortium for Pathogen Readiness.[10] That meant converting an adversarial relationship into a collaborative one, like the relationship between a primary care physician (the traditional healer) and a specialist (the Western HIV treatment doctor). It also meant helping traditional healers and HIV doctors communicate effectively with patients.

Heller sketched out a workable, collaborative narrative for the traditional healers to share with their patients: "'What we know, we've learned from our ancestors, but this is a new disease. Our ancestors didn't know this disease, and that's why we don't know how to treat this disease. So, for this disease, it's okay to go to the other doctors. But for your other conditions, you can still come back to us. For your headaches and your nausea and everything, you can still come back to us.'"[11]

Once the traditional healers started working with Western HIV doctors and were assured that they wouldn't lose their patients, the next step was teaching the healers how to perform HIV testing. This was framed as a benefit to the healers: The program awarded them a certificate and an emblemed scrub shirt upon completion. "So," said Heller, "they now practice a higher level of traditional healing, where they not only do traditional healing, but they have access to other medical methods also."[12]

Such approaches can be applied to the rollout of Covid testing, treatment, and vaccination campaigns in many countries where large segments of the population rely on traditional medicine. Addressing consumer vaccine hesitancy means getting a buy-in from local healthcare providers (and other community leaders). For widespread adoption and distribution, it's often better (and certainly faster) to collaborate with existing players than to create a new operation that competes with them. This is particularly true when existing providers of products and services have built longstanding, trusting relationships with their customers.

Convenience: Make It Easy

Although people speak of "receiving a vaccination," the experience in the US, especially in the early days of the Covid vaccination campaign, was nothing like receiving an Amazon box that magically appears at one's door. It was a lot more like a trip to the doctor in a bureaucratic healthcare system with a time-consuming, multi-step process of arranging an appointment, driving, parking, walking, and waiting, not to mention the administrative steps at check-in. Worse, vaccination could come with many added uncertainties for the customer, who wonders, "Where will I have to go? Can I find the place? What questions do I have to answer? How long will it take? Will they actually have the vaccine?" Add fear of needles and the potential for some minor side effects such as a sore arm or some flu-like symptoms, and many people formed a vast middle group between the vaccine enthusiasts and the vaccine avoiders—those willing to be vaccinated but not willing or able to put in the time and effort required.

That led many providers to come up with more convenient approaches. Prof. Gil Epstein of Bar-Ilan University outside Tel Aviv described how Israel boosted its vaccination rates by sending mobile vaccination units to parks and other outdoor gathering places on sunny weekends. "They took out an ambulance or a big car, and they brought it out and they started telling people, 'Whoever needs to be vaccinated, you can be vaccinated right here and right now.'"[13]

Similarly, in the US, some local health departments collaborated with sports teams to set up mobile walk-in clinics at local sporting events.[14] After US pharmacy chains had dealt with the crush of vaccine enthusiasts and demand slowed, they began to offer day-of and walk-up appointments.[15] And in a creative approach to combat vaccine hesitancy, a team effort involving business and community leaders, the University of Maryland and personal care brand SheaMoisture, with support from the White House, devised the "Shots at the Shop" initiative. The program combined convenience with trust by training hair stylists and barbers in Black neighborhoods across the US to advocate vaccination, provide accurate information, and administer vaccines.[16] Stephen B. Thomas, a health policy professor at the University of Maryland, summarized the approach: "Why not go where people already have trust—the barbershop and the salon?" The organizers added that they hoped they could turn this approach of working with local, trusted community partners into a national model for vaccination campaigns.[17]

Cost and Incentives: Show Me the Money

As the vaccination campaigns ramped up and supply caught up with demand, the challenge became one of convincing vaccine-hesitant people to get the jab. Governments around the world resorted to a range of incentives. Israel's approach to convince the young and vaccine-hesitant was based on offering access to social events and venues: *Green Pass* is a certificate (typically presented on a smartphone) that gives the vaccinated (and those considered immune after recovering from Covid) access to many indoor venues such as restaurants, concerts, gyms, wedding halls, and stadiums that were restricted or closed due to social distancing requirements.[18] Green Pass enabled a broader return to normal life and made vaccination the gateway to that return.[19]

Other governments implemented similar programs. At the end of July 2021, France passed a law making a health pass mandatory for indoor venues.[20] As of September 2021, a total of 129 countries had instituted such requirements.[21] New York City, on September 13, 2021, became the first US city to require proof of vaccination as a condition for indoor dining, entering movie theaters, and using gyms.[22] Some governments, such as those of China and the EU, developed vaccine passport systems that enabled travel. The US government, as well as a number of companies and universities, require proof of vaccination for entry into their facilities.[23]

However, such passport systems prompted objections in some Western countries across the political spectrum. Conservatives objected to the infringement on freedom of movement and to government tracking. They also asserted that passport systems amounted to a de facto vaccination mandate.[24] The political left voiced concerns that the passports would effectively make second-class citizens of the unvaccinated and that they would be more likely to burden minorities and the poor, who have less access to vaccines and less trust in government.[25] Moreover, in countries with limited supplies of vaccines and prioritized allocations, younger citizens who were ineligible for vaccination felt it was unfair to relax social restrictions only on the vaccinated.[26]

Some jurisdictions (and companies) offered direct financial incentives for vaccination. For example, West Virginia offered a $100 savings bond to young people (ages 16–35) to help the state reach herd immunity levels. "Our kids today probably don't really realize just how important they are in shutting this thing down," said Governor Jim Justice. "I'm trying to come up with a way that's truly going to motivate them—and us—to get over the hump."[27] West Virginia, known for an abundance of outdoor recreation opportunities, also selected lottery prizes likely to sway the hesitant, such as hunting rifles, shotguns, and lifetime licenses for hunting and fishing.[28]

Governor Mike DeWine of Ohio chose a different approach. He took inspiration from Bill Veeck, owner of several Midwestern baseball teams in the mid-20th century, who said, "To give one can of beer to a thousand people is not nearly as much fun as to give 1,000 cans of beer to one guy." Ohio announced five weekly drawings of a $1 million prize to any adult who got at least one dose of a vaccine, dubbing the incentive *Vax-a-Million*.[29] On May 26, 2021, 22-year-old Abbigail Bugenske of Silverton, Ohio, near Cincinnati, won the first $1 million prize.[30] To encourage vaccination among younger populations, the state made vaccinated 12-to-17-year-olds eligible for weekly drawings for full scholarships to any Ohio state higher-education institution.

Like other schemes, lottery campaigns drew criticism. Some denounced the use of public funds for such a cause, others questioned the effectiveness of such campaigns, while yet others considered them coercive. The campaigns, however, were widely successful. For example, vaccinations in Ohio among 16- and 17- year-olds increased by 94 percent and 65 percent among rural Ohioans of all ages.[31] Colorado, Oregon, and Massachusetts followed suit with $1 million jackpots. Other local officials used less expensive approaches ranging from free beer in Erie County, New York, to a dinner with the governor of New Jersey.

Beyond the purely humanitarian motivation to save lives, incentives for vaccination may ultimately save far more money than they cost by controlling Covid and eliminating the escalating costs and economic losses caused by the pandemic. As Ohio Governor DeWine wrote in an opinion piece in the *New York Times* on May 26, 2021, the day of the first drawing, "I thought about how much money the country had already spent fighting the virus, including millions of dollars in health-care costs, the lost productivity and the lost lives. Frankly, the lottery idea would cost a fraction of that."[32] His statement was backed by an independent scientific analysis demonstrating that while the program's cost was $5.6 million, the benefits—in hospital bill reductions alone—totaled $66 million.[33]

The strongest incentives, of course, are mandates. Pakistan, for example, is requiring all citizens to be vaccinated. Region-specific penalties for refusing the vaccine include cutting off cell phone service, withholding salaries, denying pension benefits, prohibiting travel to popular northern mountain destinations, and many others.[34]

Companies Do Their Part

Companies and other private organizations played key roles in promoting vaccine adoption. Beyond the obvious examples of retail pharmacy chains directly providing vaccination services and sports teams hosting vaccination sites at their venues, many firms drove adoption through promotions, incentives, and even mandates. These efforts focused on both consumers and workers. On one level, these efforts could be considered part of corporate social responsibility in supporting the health of workers, customers, and the community. On a deeper level, the efforts were about accelerating the reopening of the economy, ending the economic and operational uncertainties of the pandemic, and supporting a post-pandemic rebound in business.

Incentives for Customers and Staff

Companies promoted vaccination among customers, like with Krispy Kreme's offer of free doughnuts[35] or Budweiser's offer of free beer.[36] Several colleges and universities offered gift cards and other financial incentives to students. Both Lyft and Uber offered free rides to vaccination appointments.[37] Meanwhile, United Airlines gave away 30 pairs of round-trip tickets to members of its frequent-flier program who presented their vaccination records.[38]

Many companies crafted incentives and accommodations to encourage employee vaccination as well. In mid-January 2021, discount retailer Dollar General offered hourly employees a vaccination bonus equal to four hours of pay and adjusted stores' labor management systems to give those workers time off to get their shots. "We do not want our employees to have to choose between receiving a vaccine or coming to work," the company said in a press release.[39] Similarly, grocery chain Kroger offered $100 cash and other rewards to employees who got vaccinated.[40]

Companies that depend on large, in-person workforces—such as front-line retail, e-commerce fulfillment, and meatpacking—lobbied governments to prioritize their workforces for vaccine eligibility.[41] "There's no doubt that getting their employees vaccinated is going to be good for business and will be an important boost to getting the economy back on track," said Arthur Herman, senior fellow at the Hudson Institute, a think tank in Washington, D.C.[42]

Mandates for Employees

Some companies, especially those in the healthcare and long-term care sectors, issued vaccination mandates for employees. "As healthcare workers," Dr. Marc Bloom, CEO of the Houston Methodist hospital system, told employees, "we must do everything possible to keep our patients safe and at the center of everything we do."[43]

Hundreds of universities mandated student vaccination ahead of their fall 2021 terms.[44] Some, like MIT[45] and Yale,[46] went a step further and mandated vaccination for all faculty and staff as well (with limited exceptions).[47] The New York Stock Exchange requires vaccination for anybody entering its trading floor, as do other Wall Street firms, including Putnam Investments, Morgan Stanley, Jefferies, and others.[48]

On July 29, 2021, President Biden announced that all federal workers must be vaccinated or comply with new requirements, including masking, distancing, and frequent testing, among others. Several states issued a similar directive, while others limited the mandates to healthcare workers and those in long-term care facilities. And on September 9, Biden went a step further, extending the vaccination mandate to all companies with more than 100 employees.[49]

Legal Obstacles to Workplace Mandates

"Vaccination mandates are ethical," said Lawrence O. Gostin, a professor of health law at Georgetown University Law Center. "Everyone has a right to make decisions about their own health and welfare, but they don't have a right to expose other people to potentially dangerous or even lethal diseases."[50] The US Equal Employment Opportunity

Commission issued extensive guidance permitting vaccination mandates with certain provisos and requirements for reasonable employee accommodations.[51]

Issues around mandates became especially thorny in industries hit hardest by Covid. For example, Covid completely shut down the cruise industry—cruise ships being one of the early headline-making superspreader environments, where unavoidable social contact indoors fomented the spread of the virus. In order to ensure the safety (and willingness) of both passengers and crew to return to shipboard life and fun, Norwegian Cruise Line implemented a "Sail Safe" vaccination mandate that applied to employees and passengers alike.[52]

However, Florida Governor Ron DeSantis issued executive orders prohibiting the creation of vaccination passports and prohibiting companies from requiring proof of vaccination from customers.[53] They would face a $5,000 fine for each occurrence.[54] "In Florida, your personal choice regarding vaccinations will be protected, and no business or government entity will be able to deny you services based on your decision," DeSantis said. As a result, Norwegian considered suspending all Florida departures in favor of other states and Caribbean ports.[55] In July, the cruise line sued Florida's surgeon general in federal court, accusing the state of preventing it from safely resuming operations.[56] The judge sided with Norwegian and issued an injunction that prevented the enforcement of the Florida rule,[57] and other cruise lines, including Carnival and Virgin, followed suit.[58] In September, Florida started issuing a $5,000 fines to businesses that required customers or visitors to show vaccination proofs. Cruise lines, however, were exempt due to the previous court decision.[59] As demonstrated by this example, the evolving landscape of state-specific laws in the US poses challenges to companies with a national footprint.[60]

Resistance to mandates also came from employees. For example, in April 2021, when Houston Methodist hospital system mandated vaccination (with limited exceptions) as a condition for continued employment for its 26,000 staff members, a handful objected. Although more than 99 percent of the company's workers complied, 153, including 100 nurses, sued. The anti-vaccination workers lost their court case[61] and were subsequently fired.[62] Similarly, at the end of September, United Airlines began the process of terminating 593 employees who refused to comply with its vaccination mandate (out of a work force of 67,000).[63]

Side Effects

Whether customers adopt a new product depends on their confidence that the product's benefits outweigh its costs and risks. In the case of vaccines and other pharmaceuticals, clinical trials are designed to document both the benefits (e.g., avoiding disease and death) and risks of side effects or safety issues (e.g., chances of headache, muscle aches, fever, or worse). This data, collected from tens of thousands of Covid vaccine trial participants, was designed to demonstrate whether those who got the vaccine experienced fewer health issues or deaths compared to the unvaccinated (those who received a placebo). Overall, the clinical trial data established that, on average, being vaccinated was significantly safer than being unvaccinated.

That said, such trials cannot catch extremely rare side effects. For example, if a clinical trial has 10,000 people in its vaccinated group, then any side effect that occurs in fewer than 1 in 10,000 cases in the general population is unlikely to be discovered during the trial. And if, for instance, 5,000 subjects are women, with 1,000 women in each 10-year age cohort (e.g., 20–29, 30–39, etc.), then a side effect that occurs in less than 1 in 1,000 young women in a specific 10-year age cohort might not show up in the trial. Uncovering the rare side effects in subpopulations requires more data.

Data collection on rare side effects does happen steadily as people get vaccinated. National regulators create monitoring systems that gather data on anomalous medical conditions that may potentially be rare adverse side effects of vaccines. For example, the FDA maintains the Vaccine Adverse Event Reporting System (VAERS). In mid-April 2021, VAERS highlighted six cases of women between the ages of 18 and 48 who suffered severe blood clots (including one death) within a week or two of getting the Johnson & Johnson vaccine.[64] At that time, the US had already administered 6.8 million doses of the J&J vaccine. At first glance, this seemed to imply that the vaccine caused only a one-in-a-million risk of blood clots. But was that a valid conclusion, and how should regulators have responded to the data?

No Easy Options

In deciding whether to halt use of the J&J vaccine, authorities faced two problematic options. On one hand, if they halted injections of an effective vaccine while Covid continued killing 1,000 people per day in the

US—and while other vaccines were in short supply—it would surely delay millions of vaccinations and lead to even larger numbers of additional cases and deaths. In addition, news of the halt would certainly confirm the fears of the vaccine-hesitant, further reducing vaccination rates. Alternatively, if authorities continued distributing the J&J vaccine, they risked a subsequent revelation of the adverse side effect data. At that point, the media (and partisan adversaries) would surely call it a cover-up, severely damaging the already low trust in government and science in certain groups and dramatically increasing vaccine hesitancy.

On April 13, 2021, the FDA and CDC announced the existence of the six cases and recommended pausing injections of the J&J vaccine. The agencies called for a pause because VAERS (and similar systems) have limitations. The initial one-in-a-million estimate of risk may have been misleading; the true value might either be reassuringly lower or alarmingly higher. On one hand, VAERS may overestimate risks because the system does not require any proof that the vaccine caused the adverse condition. Every day, ordinary people suffer the onset of all manner of strange and serious maladies, but that does not mean these unusual medical conditions were necessarily caused by whatever unusual action the sufferer recently took, such as getting vaccinated. The logical fallacy of *post hoc, ergo propter hoc*[†] should not rule policy decisions.[65]

On the other hand, VAERS can also underestimate risks for several reasons. First, VAERS is a voluntary data collection system, not a comprehensive one that taps detailed healthcare records of every vaccinated person (a disadvantage of the US over countries with more centralized, digital healthcare systems like Israel or the UK). Thus, an unknown additional number of vaccinated patients could be suffering from some VAERS-logged possible side effect, but neither the patients nor their doctors ever saw a connection between the condition and vaccination or took the time to report it to VAERS. Second, some side effects may take time to manifest—thus, cases in VAERS may only reflect the effects of early J&J vaccinations, not all 6.8 million. Third, the simple one-in-a-million estimate for the original six cases in 6.8 million shots implies that every one of those 6.8 million vaccinated people was

---

[†] *Post hoc, ergo propter hoc* (Latin for "after this, therefore because of this") is a logical fallacy in which one event is (incorrectly) said to be the cause of a later event simply because it occurred earlier. It is a fallacy because correlation does not equal causation.

equally likely to suffer the side effect, which may be wrong. Further study may find the risk is concentrated in a subgroup.

Indeed, later studies found nine additional cases of blood clots for a total of 15 cases and three deaths among women ages 18–60.[66] Instead of a one-in-a-million risk, the additional data and analysis estimated the risk level as 12 per million in women age 30–39 and seven per million in the broader category of 18- to 49-year-old women. (The investigation also showed that the mRNA vaccines did not appear to induce this blood clot side effect despite being administered to 26 times as many people as the J&J vaccine.)[67]

Risk in the Balance: J&J Beats Covid

Finally, any newfound unusual risks of vaccination must be carefully compared to the well-documented, serious risks of remaining unvaccinated. Whereas vaccination may create some rare, one-time risks (like rare blood clots within a week or two), being unvaccinated creates a much higher ongoing risk of disease and death—and spread—for as long as Covid continues to circulate in the population.

At the time of the FDA's investigation in mid-April 2021, Covid was infecting more than 2 million unvaccinated Americans and killing over 23,000 of them every month.[68] Moreover, Covid itself caused blood clots in an astounding 20 percent of those with the disease,[69] along with a litany of other dangerous symptoms and side effects. The US was nowhere near herd immunity—some people were getting reinfected with Covid, and rising rates of mutant variants of the virus were increasing the likelihood that anyone who remained unvaccinated would face months (and possibly years) of ongoing risk of disease.

Even among young women ages 18–49, who have the highest risk of vaccine-related blood clots and much lower risks of Covid-related medical problems, the updated risk analysis showed that Covid was more dangerous than the J&J vaccine by an order of magnitude. Whereas a million vaccinated women in this age group might lead to 13 cases of blood clots and one or two deaths, a million unvaccinated women would lead to more than 600 hospitalizations, almost 130 ICU admissions, and 12 deaths.[70]

After a 10-day pause, more data collection, and much analysis, the CDC concluded that resuming the use of the J&J vaccine was safer than letting people remain unvaccinated.[71] Moreover, because the J&J vaccine only required one dose, it had a better chance to get more people fully vaccinated in comparison to the two-dose vaccines, the second dose of which some people skip. Finally, to minimize the severity of the side effects, regulators publicized the potential risk so that patients could watch for symptoms of blood clots, and they also gave doctors instructions on how to best treat them.

Importantly, regulators did not use VAERS to draw conclusions; they used it to trigger investigations to decide whether an abnormality was a coincidence or a problem of true concern. VAERS is figuratively the canary in the coal mine—a sensitive early-warning signal that *might* flag a possible problem with a vaccine. VAERS and similar systems, along with follow-up investigations, serve three purposes: (1) to catch rare but severe side effects that may change the regulatory approval status of a treatment, (2) to modulate who should receive the treatment based on the updated risk analysis, and (3) to alert doctors (and vaccine recipients) to watch for these effects so they can be quickly and correctly treated.

Note that this monitoring function of VAERS is not unlike the monitoring systems used by companies to track problems surfaced by social media reports of product defects, warranty repairs, and other service problems. News of a problem isn't always bad news if it reflects the normal rate of rare random defects, abusive customers, or other accidents. However, repeated reports of a problem necessitate an investigation. Depending on the outcome of the investigation, the company might tweak the product's manufacturing; change the design; offer a free, optional remediation for concerned customers; or issue a full recall in the case of a serious issue.

Other Responses to Side Effects

A similar issue in Europe elicited more nuanced initial reactions among European countries. By April 2021, some 222 blood clot cases and 30 deaths were reported among 34 million people who got the first dose of the AstraZeneca vaccine.[72] These cases were concentrated in younger people; as a result, the UK stopped using the AstraZeneca vaccine for people under age 30,[73] and Germany similarly restricted AstraZeneca for those under 60.[74] Australia recommended use of the AstraZeneca vaccine only for people over 50 years old and opted for the Pfizer–BioNTech vaccine for those under 50.[75] The rationale for these policies was to reduce the risk of blood clot side effects in the young by taking the AstraZeneca vaccine off the table for that age group while at the same time reducing the risk of Covid in the elderly by utilizing all available vaccine doses as soon as possible.

It's interesting to note that the two vaccines that have been linked to this kind of serious blood clotting are both based on the same vaccine technology: a non-replicating adenovirus viral vector. In July 2021, it was reported that both Johnson & Johnson and AstraZeneca were exploring vaccine reengineering in response to the rare blood clot problem. Time will tell whether Russia's Sputnik V vaccine, which is also based on adenovirus viral vector technology, has the same problem. Unfortunately, in addition to Russia, where information is tightly controlled, Sputnik V is mostly used in developing countries where data collection and reporting of side effects is not as robust as in the EU and the US.

On July 9, 2021, the FDA issued a new warning for the J&J vaccine following reports of 100 cases of Guillain-Barré syndrome (GBS), leading to one death.[76] Given that, by then, 13 million Americans had received the J&J shot, the FDA did not hold up the J&J campaign because of the rarity of the condition. GBS takes place when a person's nerve cells are attacked by the body's own immune system, leading to possible (usually temporary) paralysis.[77] The same rare side effect, at about the same rate, was observed in recipients of the H1N1 swine flu vaccine in 2009–10.[78] Note that 3,000–6,000 people in the US develop GBS every year regardless of any vaccination.

In another instance, rare cases of cardiac inflammation following mRNA vaccinations have been reported: myocarditis (inflammation of the heart muscle) and pericarditis (inflammation of the thin membrane surrounding the heart). While both conditions are relatively rare, the data indicate a statistically significant increase relative to pre-vaccination levels, mostly among young male adults.[79] In this case, the CDC did not halt the vaccination process. It only issued advisories and reports while it continued to monitor the data and investigate the issue[80] because most cases were mild and resolved within a few days.[81] In addition, patients infected by Covid-19 had nearly 16 times the risk for myocarditis compared with patients who had not been infected.[82]

The Side Effects of Side Effects

Infectious disease expert Dr. Howard Heller illustrated how mistakes in vaccination campaigns can have serious long-term consequences, as in the case of the 2016 campaign in the Philippines to vaccinate against dengue fever. Dengue fever—also known as breakbone fever—is a painful, mosquito-borne viral disease found in tropical and subtropical areas.[83] Every year it infects about 400 million people, sickens about 100 million, and kills about 22,000.[84] The virus has a unique biology with four distinct serotypes. Getting dengue of one type provides no immunity to the other three. In fact, it's the opposite: subsequent infections by other types are often worse.

After 20 years of development and $1.8 billion in costs, Sanofi introduced a dengue fever vaccine (Dengvaxia) in 2016.[85] In April 2016, Philippine President Benigno Aquino III embarked on a dengue fever vaccination initiative. The Department of Health spent $67 million on Dengvaxia and kicked off a mass immunization program with the aim of vaccinating a million students by the end of the year.[86] Heller described Aquino's push for the vaccination effort: "He had this whole big publicity campaign that he was doing this for public good. They had slogans and cartoon characters and colors and costumes. And then people started to die—especially children."[87]

The problem was that if Dengvaxia was given to someone who had never had dengue before, their risk of severe dengue became three times higher than if they had remained unvaccinated.[88] This was especially likely among children, who had previously been least likely to catch dengue. "All officials who spoke with me about the Dengvaxia campaign worried about the potential for future severe dengue cases in vaccinated persons who had never had a previous case of dengue," said Dr. Jody Lanard, a public health risk communication expert who witnessed the April 2016 rollout of Dengvaxia in Manila.[89]

The Philippines halted the campaign and ordered Sanofi to cease the sale and distribution of Dengvaxia in the country, but scandal, protests, and lawsuits ensued. Worse, the damage went beyond those who had received Dengvaxia. A year later, a second health disaster broke out when a measles epidemic hit the country. The Department of Health then, as before, embarked on a measles vaccine campaign. "Do you think the people were eager?" asked Heller. "Talk about vaccine hesitancy. They had just come through this dengue disaster, and now they're trying to convince people to get a measles vaccine, and the people wouldn't do it. So, they ended up having a lot of the people dying of measles."[90] Public distrust of vaccination in the Philippines caused more than 33,000 measles cases and 466 deaths from this obviously and safely vaccine-preventable disease.[91]

Clearly, violating people's trust has serious consequences, especially for public health. When it comes to evaluating products, people have a natural hierarchy of risk aversion and tolerance defined by whether the product is "in me" (such as medicine and food), "on me" (such as cosmetics), or "around me" (such as furniture and cars).[‡] With vaccination being as "in me" as one can get, many have deep concerns about any risks. (That aversion is no doubt many times greater when it comes to "in me" products for children.)

Whether Sanofi obscured the risk or the Philippine government ignored the risk is beside the core point about the much bigger picture of the risk management environment of vaccination. Although delaying a promising vaccine or treatment seems foolhardy, botching a public health campaign is much worse. The risks go beyond the trade-offs between deaths in the unvaccinated versus the vaccinated for any disease du jour. It is crucially important to protect the public's fragile trust in science and government, which in turn influences subsequent rates of death and disease in all future public health initiatives.

In fact, the high vaccine hesitancy among the Black community in the US[92] can be traced to the US government using Black citizens as unwitting test subjects in medical experiments. The most well-known of these are the 1932–1972 Tuskegee syphilis study[93] and the Henrietta Lacks cell studies from 1951 onward.[94] Consequently, despite Blacks dying from Covid at three times the rate of white populations, the stain of these historical crimes is one explanation for vaccine hesitancy in this population—and for a general distrust of medical and governmental institutions.

[‡] The concept was first suggested by Bill Morrissey, vice president of environmental sustainability at Clorox, in the context of environmental attitudes at the 2008 Sustainable Brands Conference.

The Never-Ending Battle for the Truth

The Covid-19 pandemic was more than just a physical virus that attacked the lungs and other organs of the human body. It also spawned a pandemic of viral memes on both sides of the debates about the best measures for tackling (or living with) the pandemic. The combination of a rapidly spreading and evolving new disease; the daring, innovative healthcare technologies; and the life-upending government policies created a cloud of unnerving unknowns from which emerged a rogues' gallery of the misinformed, the fearful, and the exploiters of chaos. The collision of complex science, frightening anecdotes, nonintuitive dynamics of pandemics, and the frustrating behaviors of fellow citizens and politicians all sowed fertile ground for confusing, conflicting, and misleading information about Covid.

Some of the disputed facts about both the pandemic and the ongoing attempts to control it did little more than provide fodder for the ephemeral theater of the 24-hour news networks, opportunist politicians, and internet flame wars. But some of it plainly affected people's behavior (e.g., unwillingness to wear masks, reduce social contact, or get vaccinated) that, in turn, has prolonged and worsened the pandemic.[95] For example, the net result of US President Donald Trump's widely publicized views on Covid was very low vaccination rates in counties that supported him in the 2020 election,[96] which have contributed to surging rates of disease in those same places in the summer of 2021. In addition, the ever-changing messages from the CDC without proper and detailed explanations, as well as the evolving safety protocols— such as requiring San Francisco marathon competitors to run wearing masks[97]—added to the confusion and skepticism about subsequent recommendations.

A Bit of a Stink About How People Think

The psychologist Daniel Kahneman is famous for his work on how people think. He disproved a core tenet of economics—that people are rational decision makers—for which he was awarded a Nobel Prize (in economics!). In his work, he explains what it means when people think they "know" something. "Knowing" means the lack of doubt in what is actually a belief, not necessarily a fact. And people develop excessive confidence in their beliefs, which they "know" to be true, by relying on the word of trusted sources. Thus, most people who believe in climate change trust the scientists who created the climate models; people who believe in their God trust their religious leaders and teachers; and many who believe that the 2020 elections in the US was fraudulent trust President Trump and his followers. For each of these groups, respective media outlets amplified the messages that each of these groups of people already believed. Invariably, the media outlets, using their own experts and communicating with their (already convinced) audiences, provided "proofs" for such knowledge.

Kahneman explained that the notion that people carefully weigh unbiased data and rational arguments to form a logical conclusion to develop knowledge is just not the case. Kahneman is quoted as saying, "When people believe a conclusion is true, they are also very likely to believe arguments that appear to support it, even when these arguments are unsound."[98] Called *confirmation bias*, this concept means that people tend to cherry-pick the subset of facts that supports the conclusion they already believe rather than carefully arrive at a conclusion that fits *all* the facts.[99]

Indeed, a May 2021 YouGov poll revealed that about 20 percent of Americans said they would not get vaccinated. Furthermore, 80 percent of them claimed that there was nothing that could convince them to change their minds.[100] They harbored no doubt at all in their positions, validating Kahneman's definition of what it means to "know" something. Worse, efforts to convince people to get vaccinated can often have the opposite effect. "People were thinking, why are they so pressed to give people incentives to take the vaccine?" said Gerard Dupree, who won Maryland's vaccination incentive lottery but has many anti-vax friends and family. "For people who have mistrust, it just makes you more suspicious."[101]

In his 2011 book *Thinking, Fast and Slow*,[102] Kahneman recounts an experiment in which he asked college students if the following chain of reasoning is true:

> All roses are flowers.
> Some flowers fade quickly.
> Therefore, some roses fade quickly.

A large majority of college students said this syllogism was valid. In fact, the argument is flawed, because it is possible that there are no roses among the flowers that fade quickly. Yet, people know that the conclusion is true and so they think the line of reasoning is valid.

Dr. Robert Kaplan, research director of Stanford's Clinical Excellence Research Center, noted that since August 2020, before any Covid-19 vaccine was available, 20 percent of the US population claimed that they would never get one, and another 15 percent said they were unlikely to. Subsequent polls revealed that these percentages did not budge as vaccines came out following successful clinical trials, and even after hundreds of millions of people were vaccinated.[103] Thus, as Kahneman argued, data cannot change somebody's firm "knowledge."

Even personal experience was sometimes not enough to convince people. Dr. Clay Dunagan, chief medical officer at BJC HealthCare in Missouri, said, "We have had patients that, even when they come down with Covid, don't believe Covid is a real thing; even when they are in the ICU and on a ventilator."[104]

Such examples explain why new data and reasoning can fail to convince people who are self-assured about their knowledge that they have any chance of being wrong. While this framework of human thought can explain most of the vaccine hesitancy and anti-vaccine positions, there are many other contributing factors, none of which are new.

The History of Hysteria

Opponents of vaccination—those so-called anti-vaxxers—date to the beginning of vaccinations shortly after Dr. Jenner's 1796 experiments showed that inoculation with mild cowpox virus created immunity to the deadly and disfiguring smallpox virus. An 1802 satirical cartoon suggested that the cowpox vaccine would cause people's bodies to erupt with cow heads and horns.[105] Public health regulations requiring vaccination engendered protests and spawned anti-vaccination leagues in the mid-19th century.[106]

Just like any other "knowledge," pro- and anti-vaccination sentiments are a function of whom and what people trust. Vaccine proponents trust medical advances, technology, their doctors, and pro-vaccine politicians with whom they agree politically. Anti-vaxxers trust different people, including their like-minded neighbors, religious leaders, or politicians with whom they agree politically. Some are part of naturalistic movements (opposing injection of artificial substances) and take their cues from the leaders and healthcare providers supporting such views, while others hold strong beliefs in personal liberty above all and are not willing to get vaccinated simply because the government recommends it.

The fact that a large portion of individuals in developed countries are anti-vaxxers, despite all the vast data about the safety and efficacy of most Covid vaccines, is a testament to Kahneman's view of people's decision-making processes. Anti-vaxxers seek support for their "knowledge" from other sources. Interestingly, the highest percentage of people who do not believe that vaccines are safe is in Europe. In a 2019 Gallup poll, a third of the French said they did not believe that vaccines are safe,[107] dovetailing with data showing that more than half of French people distrusted science and technology. Similarly, half of Ukrainians believed that vaccines do not work. It is no surprise, then, that measles cases tripled across Europe in 2020, largely driven by 54,000 cases in Ukraine.[108]

While objective evidence suggests that the Covid vaccines are safe and effective, multiple elements contribute to the generation and propagation of "knowledge" that contradicts the science. First, the complexity of medical issues in the face of fundamental uncertainty leaves much room for different interpretations of the data, models, and conclusions. Second, the media, especially today's social media platforms, make the spread of different points of view very easy. Third, warped incentives mean that the media and other providers and distributors of information are rewarded for information that is scary and alarming, rather than sober, accurate, and rooted in facts. Fourth, and worst of all, mistrust in institutions and governments (some of which may be well-deserved) creates an obstacle to effective official communication—including data-supported facts. Distrust in the messenger prevents acceptance of the message, no matter how well-reasoned.

(Mis)Understanding

The Covid-19 pandemic sparked a lot of discussion, armchair analysis, media punditry, and vitriolic debate about the disease, treatments, preventative measures, vaccines, and policies. "Covid has turned us all into amateur scientists," said Talya Miron-Shatz, an associate professor and expert in medical decision-making at Ono Academic College in central Israel. "We are all looking at data, but most people are not scientists."[109] The resulting uninformed proclamations led to unintentional misunderstandings on the part of citizens, journalists, politicians, and even some scientists about what the data actually meant.

The example described earlier of the risks of blood clots with some vaccines illustrates some of the many challenges to correctly understanding and interpreting the data surrounding an important issue. The first are biases in the data that may be caused by an emphasis on anecdotes, underreporting, overreporting, wishful thinking, biased sample, or biased observers. For example, a statistical analysis of fish caught in a given lake concluded that there are no fish smaller than the holes in the net. Such results ignores the biased sample which includes only fish that were caught. The second are challenges in understanding the meaning of statistical results. For example, claiming that a statistician was drowned when crossing a river whose average depth is only 50 centimeters (20 inches) sounds implausible if you ignore the fact that the "average" value says nothing about the deep end of the river. The third is understanding the limits of any kind of data, such as the statistical chances that something looks true due only to blind randomness. Finally, there is the widespread confusion of correlation with causality. While it is true that the damage from a fire is directly correlated with the number of fire trucks at the scene, the fire trucks do not cause the damage.

Consequently, investigators (and commentators) face the problem of identifying a plausible cause of any adverse health events, analyzing their implications, and readjusting public health policies to fit the new data. Such analysis either strengthens confidence in the initial suspicion (e.g., the vaccine caused the blood clots), dispels the concern by showing that an apparent side effect actually has causes unrelated to the vaccination, or establishes that the occurrences are so rare that the dangers from contracting and dying from Covid far outweigh the risk of the side effect (in this example, the clots). The marshaling and analyzing of data helps determine if the vaccine really does have a side effect and, if so, what to do about it. All of these issues compound the need for dedicated and specialized experts who understand the data, the math, the demographics, and the medical context. Thus, trust in such experts is crucial.

The scientific community has developed certain processes to reduce the possibility of trust-busting false scientific conclusions and results. These processes include peer review of scientific publications, requiring scientific experiments to be repeatable by others, careful experimental designs, and double-blind experiments (in which neither the subject nor the investigator knows if the subject got the treatment being tested or a placebo). For products that clear all these hurdles, governments have systems to monitor rare side effects, be they defects in automobiles or other products (leading to recalls) or unexpected adverse reactions to vaccines or other medications. However, trust is hard to foster and easy to lose. That's especially true when dealing with complex new phenomena, where there may be contradictory, unvetted "knowledge" sources—with easy access to information dissemination channels—that can chip away at trust.

Perhaps the deepest source of misunderstanding, and even mistrust, comes from the unavoidable chaos of the scientific process itself. When a new threatening phenomenon such as Covid appears, no one really knows anything about it because the scant data that first appears is fragmentary (e.g., a few cases of a mysterious disease), often collected under non-ideal conditions (e.g., potentially biased anecdotes), and in many cases collected and analyzed by parties that have incentives to hide the facts or tout their role in a potential discovery. Although science seems to be a set of absolutely true facts supported by evidence, it is, instead, a process for developing plausible potential hypotheses (from theories and initial data) and then gradually testing them to determine which are true, which are false, and which still belong in the "we don't yet know" column.

However, right from the beginning of a crisis like a pandemic, government leaders and the public need to make potentially life-altering decisions. Therefore, both politicians and the public demands prompt answers about existential threats like Covid. The result is that, in the beginning, many different scientists announce many different conclusions and recommendations based on the varying (and limited) bits of data and each scientist's preexisting knowledge and beliefs. Unfortunately, many if not most of those early conclusions will, in time, prove false as scientists collect more and better data. "You can be wrong if you're dealing with information that is evolving," said Anthony Fauci on the New York Times's podcast Sway.[110] That gulf between what science can do and what people expect or wish it could do creates fertile grounds for mistrust.

(Mis)Information Distribution

Rumor mills and grapevine gossip may be as old as humanity, but the internet converted the dissemination of misinformation from bounded, over-the-fence whispering into unbounded, around-the-world shouting. The birth of social media platforms created zero-cost distribution networks for anyone with any sort of belief and a way with words to attract thousands or even millions of followers. For example, an analysis of online messages found that a mere twelve leading online anti-vaxxers with a combined total of 59 million followers generated nearly two-thirds of all anti-vax social media messages across all major platforms.[111]

In a plea to combat misinformation, attorneys general from a dozen US states sent an open letter to the CEOs of Twitter and Facebook, saying, "As Attorneys General committed to protecting the safety and well-being of the residents of our states, we write to express our concern about the use of your platforms to spread fraudulent information about coronavirus vaccines and to seek your cooperation in curtailing the dissemination of such information." The letter added, "Unfortunately, misinformation disseminated via your platforms has increased vaccine hesitancy, which will slow economic recovery and, more importantly, ultimately cause even more unnecessary deaths."[112]

Social media platforms have taken some steps to reduce anti-vax information. Facebook claimed it had removed some 2 million examples of false information related to Covid from Facebook and Instagram. "We've also labeled more than 167 million pieces of Covid-19 content rated false by our fact checking partners, and now are rolling out labels to any post that discusses vaccines," Facebook spokesperson Dani Lever said in a statement.[113] Facebook also said it was working to improve proliferation of factual information. "Since research shows that the best way to combat vaccine hesitancy is to connect people to reliable information from health experts, we've connected over 2 billion people to resources from health authorities."[114] Yet, as a series of articles in the *Wall Street Journal* revealed, Facebook fell woefully short on many of its proclamations,[115] including CEO Mark Zuckerberg's push to help the US vaccination campaign.[116]

Despite the efforts of all the main platforms, false information keeps popping up. Andy Pattison, manager of digital solutions for the WHO, said that anti-vaxxers "learn the rules, and they dance right on the edge, all the time... It's a very fine line between freedom of speech and eroding science."[117] Just as dangerous variants of the Covid virus spread faster, dangerous variants of ideas about Covid spread faster too. Bad news travels fast.[118] As the great Mark Twain said, "A lie can travel half way around the world while the truth is putting on its shoes." Indeed, an MIT researcher found that false information does travel faster, deeper, and more broadly on Twitter than real news.[119]

Incentives for Deception

The old newspaper adage "If it bleeds, it leads" encapsulates the sad truth that stories of disasters and pending calamities that evoke fear help sell media. Similarly, social media and internet media have financial models based on customer activity, such as engagement and click-throughs, which are reinforced by provocative messages leading to "doomscrolling." In the never-ending search for sales, high ratings, clicks, and screen time, media organizations have incentives to overemphasize rare but terrifying defects or flaws, or to dig up the most lurid anecdotes. As such, the various media entities profit from fearmongering, even if the information is flat-out false.

Likewise, politicians have strong incentives to spin the truth if it suits their agenda and leads to continued political power. For example, during the first year of the pandemic, politicians in many countries hid evidence of the disease or belittled its danger.[120] The Chinese suppressed information about the pandemic and delayed the isolation of Wuhan, all in order to support the public image of China as a stable power. Similarly, Iran's leaders hid the existence of Covid and the growing numbers of Covid deaths from the public in order to minimize public perception of the depth of the crisis.[121] Mexico's President Andrés Manuel Lopez Obrador told his citizens in speech after speech that they shouldn't fear Covid-19 and should go shopping.[122] Politicians who saw Covid as a threat to their power had incentives to downplay the pandemic. Brazil's President Jair Bolsonaro claimed that the disease "would be like a little flu or a little cold,"[123] echoing similar claims by US President Donald Trump.[124] Spain's Prime Minister Pedro Sánchez led a newly formed, fragile minority government and was loath to anger citizens with unpopular restrictions. Consequently, he allowed attendance at soccer games and rallies in Madrid, leading to the world's fourth-largest outbreak.[125]

Finally, the vaccine makers themselves have financial, professional, and personal incentives to highlight the advantages of their vaccines and obscure their disadvantages. They have a natural incentive to publicize the positive data they expected to get and be skeptical of or explain away any negative data. Although modern, double-blind clinical trial protocols can employ a range of fraud-prevention measures,[126] problems can still occur. A meta-analysis suggests that about 2 percent of scientists have fabricated, falsified, or modified data or results at least once.[127]

The Political Point of View

Political scientists have argued that political leaders come to power with their own personal or ideological agendas rooted in their respective views of the world. Any issue that contradicts or falls outside those views may be actively disregarded or take a back seat, regardless of the importance of the issue or the validity of the data supporting it (something that Kahneman would probably agree with).[128] For example, US President Ronald Reagan wanted nothing to do with the HIV/AIDS crisis, barely acknowledging it. He saw it as a crisis affecting gay men—not a concern for him or his religious backers.[129] By contrast, US President George W. Bush, also a Republican, rose to power on the back of "compassionate conservatism." He launched a major, well-funded program to combat the HIV epidemic in Africa, seeing it as the duty of a US president to alleviate such suffering.[130]

Of course, Republican politicians do not have a monopoly on political points of view that distort what should be data-driven decisions. For instance, President Joe Biden's team shut down an effort launched late in the Trump administration to prove the coronavirus originated in China's Wuhan lab. The administration's view was that Trump had been trying to shift the blame for the pandemic to China to distract the attention of the media and rally his nationalistic, anti-Chinese base. Later, Biden directed the intelligence community to reopen the investigation.[131] In addition, the Biden administration has also played down the crisis of asylum seekers at the US–Mexico border, its role in the execution of the withdrawal from Afghanistan, and many other unflattering matters.

Geopolitical Pandemic Propaganda

Governments also have strategic incentives to play the misinformation game—creating and disseminating propaganda to undermine economic and geopolitical rivals. For example, four publications associated with the Russian government attacked Western vaccines, trying to instill fear about their efficacy and side effects.[132] Similarly, a pro-Chinese propaganda network of thousands of fake YouTube, Twitter, and Facebook accounts sowed doubt about the safety of the Pfizer–BioNTech vaccine and other US measures against Covid.[133] According to a European Council task force report on disinformation, "Both Chinese official channels and pro-Kremlin media have amplified content on alleged side effects of the Western vaccines, misrepresenting and sensationalizing international media reports and associating deaths with the Pfizer–BioNTech vaccine in Norway, Spain and elsewhere."[134]

The European External Action Service, an arm of the EU diplomatic service, accused China and Russia of using disinformation in their vaccine campaigns. It said, "so-called 'vaccine diplomacy' follows a zero-sum game logic and is combined with disinformation and manipulation efforts to undermine trust in Western-made vaccines, EU institutions, and Western/European vaccination strategies."[135] US State Department researchers and officials said that Russian news outlets connected to election disinformation campaigns in the United States were now targeting Latin American and African countries with disinformation campaigns to attack American coronavirus vaccines and boost trust in the Russian vaccine.[136]

Mistrust in Information Suppliers

Around the world, trust altogether has been declining, according to the public relations firm Edelman, which publishes an annual "trust barometer."[137] Edelman's 2021 barometer uncovered a marked decline in trust around the world in all categories of societal leaders: government leaders, CEOs, journalists, and even religious leaders. Yet business was the most trusted source—it is perceived as both ethical and competent—followed by NGOs, government, and the media. The pandemic drove trust in all news sources to record lows, with social media being the least trusted. The widespread erosion of trust made the system unable to confront the rampant "infodemic" of false and biased messaging.

"Vaccine hesitancy has less to do with misunderstanding the science and more to do with general mistrust of scientific institutions and government," said Maya Goldenberg, a philosophy expert at the University of Guelph, Ontario, who studies this phenomenon.[138] The net result of both partisan and malicious false information is an overall erosion of trust, as documented by the Edelman Trust Barometer mentioned above. Dr. Fauci noted that "there is a general anti-science, anti-authority, anti-vaccine feeling among some people in this country—an alarmingly large percentage of people, relatively speaking."[139]

A week before his election to the US presidency, Joe Biden tweeted, "I believe in science. Donald Trump doesn't. It's that simple, folks."[140] If only it were that simple. British science writer Donald Ridley contends that the pandemic highlighted the disconnect between science as an idealized philosophy and science as a human institution. The former represents the "primacy of rational and objective reasoning." The latter, like all human institutions, is subject to political dogma, human needs and wants, and can be erratic and prone to failures.[141]

While Ridley used the example of the Chinese Communist Party's total censorship on scientific publication, the problem, in less extreme forms, exists in every scientific institution. For example, the pressure to "publish or perish" in universities leads to many subpar scientific publications and pressure to bias or even fabricate results. Peer review— intended to filter out low-quality research—also stifles new ideas if they run contrary to the theories of venerated peers. Thus, sometimes science only advances "one funeral at a time."[142] In addition, peer pressure to avoid controversial opinions under the threat of "cancel culture" can inhibit open debate, which is crucial for scientific inquiry and arriving at a knowledge and truth that are not a byproduct of sacred notions that may or may not be truthful.

The unfortunate result can be public mistrust of messages coming from the scientific establishment. Major societal challenges, such as pandemics, climate change, or economic inequality, often involve complex phenomena with counterintuitive or hard-to-explain effects. In many cases, simple corrective initiatives, which seem intuitively "right," only exacerbate the problems. Having to study and then justify nuanced and counterintuitive courses of action is often politically impossible. In addition, the media in the age of the internet has become politicized, opinionated, and simply uneducated about most scientific subject matters. The public, who consume the output of their chosen media outlets, is left to sift between several competing "scientific truths" with no clear way to decipher what is fact-based, what is conjecture, or what may be simply figments of a journalist's, blogger's, or tweeter's imagination.

Battling Misinformation

The Israeli government fought misinformation and disinformation in its efforts to quickly vaccinate the country, bring infection rates down, and reopen the economy. In particular, as mentioned earlier, it worked to dispel false rumors that fomented vaccine hesitancy among the ultra-Orthodox, who tend to have large families, lead highly social lives, and distrust modern technologies. Health authorities worked both from the top down and from the bottom up to build trust and stamp out misinformation.

On the top-down end, the Israeli Ministry of Health worked with influential ultra-Orthodox rabbis to explain the science behind the vaccines and debunk falsehoods about risks such as infertility. Avi Blumenthal, an ultra-Orthodox public relations consultant, told NPR, "The rabbis needed to know two things, really: (a) effectiveness and (b) safety."[143] The result was that the religious leaders changed their recommendations to their congregations from "let's wait" to "let's get vaccinated."

On the bottom-up end, the government tried to snuff out sources of false information circulating in the ultra-Orthodox community. Although the ultra-Orthodox eschew technology and do not use social media, surf the internet, or watch TV, they do get information (and misinformation) from paper posters in public spaces and an array of telephone hotlines offered on "kosher" cellphones. For example, in the Haredi neighborhood of Bnei Brak, next to Tel Aviv, daily street posters appeared at one point claiming that the vaccinations were part of a dark government conspiracy and led to death. Very quickly, these anti-vax posters would be covered by pro-vax government posters quoting many religious leaders, with their pictures, who encouraged vaccinations. "They'd put one up, we'd put one up on top. They'd put one, we'd put one. It became a war," Blumenthal said. "They tired out."[144]

The kosher cellphone hotlines make money from callers through a revenue-sharing model with the caller's phone's carrier. To combat misinformation, an influential council of rabbis put the anti-vaccine hotlines on a telephone blacklist of numbers declared unsuitable for the devout. This both cut off the distribution of the false, fear-inducing information and killed the financial underpinnings of these groups.[145]

To address internet sources of false anti-vaccine information circulating among the secular population in Israel, the Ministry of Health expanded its digital task force to conduct daily monitoring of popular social media platforms in five languages and to combat anti-vax misinformation before it could spread.[146] When a pregnant woman who had not yet gotten vaccinated caught Covid, killing both her and her baby, the anti-vaxxers spun a false version of her story and publicized it. "They were claiming that my wife was vaccinated and therefore she died," said the bereaved husband. "It was like they were twisting a knife in our stomach."[147] The family, national media, and the government subsequently launched a campaign to set the story straight and turn it into an example of the horrible risks of not getting vaccinated.

Combating misinformation is an ongoing job. When applied continuously, it is like the immune system in having to be ever vigilant against an onslaught of intruders and infections. The "system" must recognize false information and react. And, as with the battle against mutations of Covid, information immune systems need to track new adaptations created by providers of false information, such as by obscuring their messages from misinformation-filtering algorithms by using new misspellings like "vacseen."

The Flaws in the Vaccination Fears

Anti-vaxxers often claim that the vaccines are experimental drugs and that the full set of side effects of these newly invented vaccines are not yet known. Technically, they are right, although the history of vaccines shows that virtually all side effects happen in the first two months.[148] Only time and science will reveal the long-term effects of the Covid vaccines. Anti-vaxxers use this fact to create fear of the unknown and to argue against vaccination.

However, this line of argument ignores two important facts. The first is that the full set of long-term side effects of Covid itself are also not yet known (and the problem of "long Covid" exemplifies how serious illnesses do have a very bad history of long-term side effects). If the vaccines are an experimental cure, then Covid is an experimental disease. The second fact is that all of the evidence to date strongly shows that the short-term effects of Covid (e.g., illness, hospitalization, disability, and death) are both orders of magnitude worse and orders of magnitude more common than the short-term side effects of vaccination (even among populations who are most susceptible to the side effects).

From Hesitancy to Immunity

One can hardly blame authorities for thinking the anti-vax problem is practically impossible to overcome. In some cases, however, government policies did help. Recall that France had a very high vaccine hesitancy rate in 2019. Yet, a combination of government pressure (in the form of mandates) and inducement (in the form of a vaccination pass required for a variety of activities, from traveling to sitting in a café) resulted in a very high vaccination rate. About 88 percent of people over age 12 in France had received at least one shot by September 2021.[149]

Government pressure, however, is not always necessary. As it turns out, people may find the path to the truth on their own, especially when the anti-vax campaigns are not egged on by some of the phenomena mentioned above. Throughout 2021, vaccine hesitancy around the world has been falling steadily, from 45 percent in January to just 20 percent in late June.[150]

Two perceptual phenomena help explain such dips. First, when the number of Covid cases spiked, the rate of vaccination increased dramatically in some countries. For example, Taiwan had been very successful in keeping the pandemic at bay through May 2021, with a total of only 12 deaths since the onset of the pandemic. Its vaccine hesitancy rate was 66 percent until then. However, between mid-May and mid-July, Covid broke through the country's containment processes and more than 700 people died. When cases and deaths surged, hesitancy plummeted to 27 percent. The same was most likely the case in Saudi Arabia.[151]

Second, while increasing perceptions of the dangers of Covid intuitively suggest a cause of decreasing anti-vax sentiments, decreasing perceptions of the dangers of vaccines has helped as well. In countries where vaccination rates were high, vaccine hesitancy also declined. For example, Singapore had also contained the pandemic well; by July 2021, it had a total of only 36 deaths.[152] As Singapore's vaccination campaign gained speed (by July 12, more than 70 percent of Singaporeans had received at least one vaccine dose), vaccine hesitancy fell from 53 percent in December 2020 to 10 percent in June 2021. Naturally, one driving force of vaccine hesitancy in these countries was the fear of the unknown effects of vaccines. As more people "survived" the vaccine, others felt better about it.

5. The Future

Widespread vaccination seems to promise an end to the tragic scourge of Covid on people's lives and livelihoods. In theory, a sufficient level of vaccination (plus immunity developed through natural infection) can provide a path to herd immunity. The concept of herd immunity originated during the epidemic of "contagious abortion" in cattle and sheep in the US at the beginning of the 20th century. In a 1916 article in the *Journal of the American Veterinary Medical Association,* George Potter and Adolph Eichhorn wrote: "Abortion disease may be likened to a fire, which, if new fuel is not constantly added, soon dies down. Herd immunity is developed, therefore, by retaining the immune cows, raising the calves, and avoiding the introduction of foreign cattle."[1] Herd immunity provides an indirect protection from infectious disease when enough of a population has become immune to an infection, whether through vaccination or previous infections, thereby reducing the likelihood of infection for individuals who lack immunity. With herd immunity for Covid in a local population, some will remain susceptible—such as the unvaccinated, babies, and the immunocompromised[2]—but their risk will diminish.

In the spring of 2021, this post-Covid future seemed well on its way. Vaccine production was rising, vaccination campaigns were accelerating, and new Covid cases were declining precipitously in communities with high rates of vaccinated people. For example, Israel was projected to reach herd immunity by March 2021[3] and likely achieved it by April or May. By June 1, with 80 percent of adults vaccinated and infection rates falling, Israel lifted some of its remaining pandemic restrictions. Dr. Eyal Zimlichman, deputy director general at Sheba Medical Center, Israel's largest hospital, said, "This is probably the end of Covid in Israel, at least in terms of the current strains that we know." And, he affirmed, "We've obviously reached herd immunity."[4]

Mutants Versus Boosters

Progress toward vanquishing the virus ground to a halt, and then reversed, with the rise of the so-called Delta, or B.1.617.2, variant of Covid, first identified in India. This lineage of the virus had accumulated 15 mutations in its spike protein, which increased its transmissibility as well as its virulence, even in immunized people.[5] Dr. Rochelle Walensky, director of the CDC, told reporters, "It [the Delta variant] is one of the most infectious respiratory viruses we know of, and that I have seen in my 20-year career."[6]

These enhanced virus attributes are a significant—and possibly insurmountable—setback in the quest to reach herd immunity. They demand an even higher percentage of vaccination and more effective vaccines to reach that goal. In addition, such mutations sparked greater numbers of cases, hospitalizations, and deaths. By August 2021, US states with inadequate vaccination rates and pandemic response measures, such as Florida, were setting new records for cases and hospitalizations.[7] Even Israel—with one of the highest full-vaccination rates in the world—saw an increase from fewer than two dozen new cases per day at the end of May 2021 to more than 10,000 per day by early September.[8]

The Making of a Mutant

As humanity adapts to Covid and tries to defeat the pathogen, the rapidly multiplying permutations of the SARS-CoV-2 virus continue to hinder humanity's efforts with their own adaptations. While people make use of their laboratories, supply chains, and governments to develop, manufacture, distribute, and fund measures against the virus, the virus counters those measures by employing the Darwinian roulette game of genetic mutations and natural selection, using the growing numbers of replicated virions[*] manufactured by the victims themselves.

When the virus infects a human being, it converts the person's cells into "bio-factories" that churn out more copies of the virus. These copies leave the infected cell to float inside the victim's body where they either find another cell to infect, are attacked by the person's immune system, or are expelled in droplets and aerosols into the air. In severe cases of Covid, the copies of the virus quickly infect more of that

---

[*] A *virion* is the complete, infectious virus particle outside a host cell.

person's cells, creating more bio-factories and making the person sicker. As the numbers of infected cells rise and production of the virus gets into full swing, the person sheds billions of viral particles (virions) in respiratory droplets and aerosols that float in the air. Other people near the infected person may be exposed, and if they are susceptible, they too will be infected.

Most copies of the virus in an infected person are identical to their parent virus. However, sometimes the hijacked cellular machinery makes a minor mistake in copying the virus's genetic code, creating a mutation that is passed on to that copy's offspring. Many mutations have no effect, some damage the copy's ability to continue infecting other cells and other people, but a few help the virus infect more cells, become aerosolized and shed into the world, or evade the immune system. As Covid spreads, new lineages of the virus acquire and accumulate mutations to create variants of the virus with different disease-related properties.[9] In short, as long as people are catching and spreading Covid, the virus has a chance to become ever more dangerous.

The Rise of "Variants of Concern"

As Covid spread around the world, labs in different countries sequenced samples of the virus and shared that data (some 1.2 million sequences by May 2021)[10] to trace the emergence and spread of mutations (including nearly 4,000 variants of the original virus, some with more than a dozen accumulated mutations). Most genetic mutations have little impact and only serve to enable tracing the vagaries of human-to-human contact and travel patterns that spread each lineage of contagion around the world.[11] However, a few mutations become what the WHO and the CDC call "variants of concern"[12] for any of several reasons.

Some variants appear to be more dangerous, driving increased rates of hospitalization and death. For example, some variants don't respond to treatment with currently available front-line antiviral drugs.[13] The original Covid variant primarily hit the elderly and (mostly) spared the young, but some of the newer variants (e.g., the Alpha variant) seem to increase the chance of severe illness in the (unvaccinated) young[14] (akin to the so-called 1918 Spanish flu, which was most severe and most often fatal in young adults). These more virulent variants increase the sad toll of the dead and the severely affected, but they have the ironic side effect of making individuals and governments more likely to act against the disease.

For example, Asian countries that had been very successful in controlling the spread of the virus in 2020 were vaccination laggards in 2021. Vaccine acceptance rates were falling during the first part of 2021 in Australia, South Korea, and Japan. Low positivity rates and normal life resuming in these countries drove people to avoid or delay vaccination.[15] Australia peaked at 600 Covid cases per day in August 2020, shut down, contained the virus, and saw low daily new case rates (often in the single digits) for almost a year, but then infections surged back to several hundred a day in August 2021.[16] The country was caught off guard without enough vaccine doses; by mid-August 2021, less than 20 percent of its population had been vaccinated. Similarly, not until June 2021 did South Korea start shoring up its vaccination rate, projecting to vaccinate 70 percent of its population by the third quarter 2021.[17]

Notably dangerous are variants that increase the infectiousness of the virus so that it spreads more quickly and to more people.[18] Some variants (e.g., the Delta variant) are less detectable by some widely used Covid testing methods, which means that infected people are less likely to quarantine and more likely to spread the variant.[19] More infectious mutations directly boost R_0, the number of susceptible people that are infected by each infectious person. Such variants exacerbate surges in infections, make it harder to contain the pandemic, and require even higher vaccination rates and more stringent social distancing and masking protocols to counter the virus. In fact, after a successful early vaccination campaign in the UK (by the end of June 2021, about 50 percent of the population had been vaccinated), a sharp rise in cases—due to the prevalence of the Delta variant—prompted UK Prime Minister Boris Johnson's government to delay the opening of the economy.[20]

As mentioned above, more transmissible variants also increase the herd immunity threshold—requiring a greater percentage of the population to be infected with Covid or get vaccinated with more effective vaccines to reach herd immunity. Epidemiologist Dr. George Rutherford of the University of California, San Francisco, estimated that conventional coronavirus strains required 71 percent of the population to be immune (via vaccination or natural infection) to reach herd immunity.[21] In contrast, he estimated that the Delta variant's increased infectiousness called for an 84 percent level of immunity.

Variants of the greatest concern are those that can evade the immune systems of the vaccinated or previously infected. Mutations in the spike protein can make the virus less recognizable by the immune system; the spike protein of the variant may look different enough that the antibodies to the original Covid virus don't bind to it as well.[22] For example, the Brazilian city of Manaus had, in theory, reached herd immunity with about 75 percent of the population having antibodies to the original Covid,[23] but then the P.1 (Gamma) variant arose and began reinfecting people. Similarly, South Africa had to suspend use of the Oxford–AstraZeneca vaccine after data showed that it was less effective against the B.1.351 (Beta) variant that was first detected in that country.[24] At the very least, such variants reduce the effectiveness of the existing vaccines, requiring even higher vaccination rates for herd immunity. At the very worst, variants that look different enough from the original strains could reset progress on vaccination back to zero, requiring both the development of new vaccines and entirely new mass vaccination campaigns.

Vax to the Max

While the Covid virus attempts to evolve in order to infect the vaccinated, vaccine makers are crafting strategies to ensure continued immunity. The easiest strategy may be periodic booster shots of the existing vaccines to increase the immune system's response to Covid, including the variants. Data from Pfizer shows that the efficacy of the Pfizer–BioNTech shot declines by 6 percent every two months.[25]

On July 11, 2021, Israel announced a booster shot campaign for adults with impaired immune systems, giving them a third dose of the Pfizer–BioNTech vaccine.[26] By the end of that month, Israel started giving booster shots to everyone over 60 years old,[27] and by the end of August, Israel lowered the booster shot age limit to 12 years.[28] Britain also began a booster campaign for the elderly and other vulnerable populations in September 2021.[29] France announced its plan to give booster vaccines to vulnerable individuals (including patients in elderly care homes, people over 75, and people with comorbidities),[30] as did Germany.[31]

Although US authorities claimed in early August 2021 that a booster vaccine was not yet necessary, some Americans were getting a booster shot on their own. They would walk into a pharmacy or a retail outlet and ask to be vaccinated without disclosing their vaccination status, and because the US lacks a central immunization database, nobody could tell that they had already been fully vaccinated.[32] Recognizing that this phenomenon is happening, the CDC has asked people to report any safety issues if they receive the unauthorized booster (even though it was not recommended) as part of the CDC's research into booster vaccines.[33] On August 13, the US authorized a booster for an estimated 10 million Americans with compromised immune systems.[34] The White House chief medical advisor, Dr. Anthony Fauci, has said that everyone will need a booster shot at some point in the future.[35] Indeed, on August 18, the White House announced plans to administer booster shots for most Americans beginning the following month.[36] The plan still needed approval from the FDA, which supported the plan, and a CDC advisory panel, which approved the shots only for elderly and immunocompromised individuals. But in a highly unusual move, the head of the CDC, Dr. Rochelle Walensky, overruled the CDC advisory panel on September 24 and expanded eligibility for booster shots to a wide array of workers across the US.[37]

Scientists know that antibody levels tend to decline over time,[38] and they soon learned that some Covid variants aren't as quickly neutralized by existing antibodies produced by previous infections or vaccinations.[39] Boosting antibody levels can help counter any reduction in vaccine effectiveness. The question is *when* to provide a booster, because scientists don't know how specific levels of antibodies correlate with specific levels of susceptibility to catching the virus, becoming symptomatic, requiring hospitalization, or dying. "As we know, Covid is not going to go away anytime soon, and we know that the antibodies decrease over time, so that a boost will be needed at some juncture. I can't predict when," said John Beigel, associate director for clinical research in the Division of Microbiology and Infectious Diseases at the US National Institute of Allergy and Infectious Diseases (NIAID).[40]

The optimization of any booster campaign—who should get booster shots, when should they get them, which vaccine should they get—is a question for science that will take time to answer. "We're entering a data-poor zone here," said Barton Haynes, a professor of immunology at the Duke University School of Medicine.[41] For example, should people get boosters of the same vaccine they had originally or a different vaccine? The former strategy might reinforce the existing immunity, and the latter might teach the immune system slightly new tricks.[42] "People just have to recognize the limitations of the data we have right now, and the critical need to generate the data to inform the decisions that matter," Beigel said. "We assume that it would be okay to give a boost with any other vaccine, but we want to make sure."[43]

Finally, an ethical question is whether any resources should be devoted to booster shots while the majority of the world remains largely unvaccinated.[44] As a matter of personal safety, many may want to increase their protection against Covid from, say, 80 percent to 95 percent. However, as a matter of public health, those same doses would make a bigger difference helping the unvaccinated get from 0 percent to 80 or 90 percent protection. In fact, on August 4, 2021, the WHO called for a halt on booster shots in order to tackle vaccine shortages in the developing world,[45] a call that was rebuffed by the US as a false dichotomy.[46] Citing United Nations data, President Biden stated that the US had donated more Covid-19 vaccine doses to nations in need than all other donor countries combined.[47]

Although factories may be stretched to the limit in producing the existing Covid vaccines, vaccine scientists continue to work on new and better vaccines. Moderna's CEO, Stéphane Bancel, said, "Leveraging the flexibility of our mRNA platform, we are moving quickly to test updates to the vaccines that address emerging variants of the virus in the clinic."[48] For example, Moderna designed a new vaccine for the Beta variant and began clinical trials in March 2021.[49] Marcello Damiani, chief digital and operational excellence officer at Moderna, explained that the trials first reporting out in May tested three possible boosters: "One is the ancestral type, one is the one with the Beta variant, and one is a combination of the two. Additional studies looking at variations of the booster strategy have been initiated since."[50]

Initial data from the Moderna clinical trials showed that a half-dose of a vaccine tailored to the virus variant, given to people previously fully vaccinated, boosted antibody response against the ancestral variant as well as others.[51] "We hope to demonstrate that booster doses, if necessary, can be done at lower dose levels, which will allow us to provide many more doses to the global community in late 2021 and 2022 if necessary," said Bancel. Part of Moderna's latest effort tests a multivalent vaccine: one that includes mRNA for more than one variant and that can train the immune system to recognize several different variants.[52]

Variants: The More, The Scarier

The longer-term challenge is that more variants will keep on coming. The already named "Delta plus" variant has all the mutations of Delta, plus a mutation found in the Gamma variant that may further help the virus infect the supposedly immune.[53] "Mutations are natural survival strategies of viruses, and SARS-CoV-2 variants are going to continue to occur throughout the world, and their number is likely to increase in the future," said Thorsten Schüller, vice president of communications at German biopharmaceutical company CureVac, which is also developing mRNA vaccines. "There will be an enormous need for vaccines in general and particularly for multivalent vaccines."[54]

Annual booster vaccination programs might entail a vaccine supply chain that provides billions of doses per year. Manufacturing and distributing another shot to the whole world sounds daunting in the face of the slow rollout of the first doses. However, while the vaccine makers have struggled to satisfy initial demand in the short term, they have, in the process, built tremendous production capacities that are likely to total many billions of doses per year in the long term. Any subsequent vaccine that uses that same technological platform (e.g., another mRNA vaccine) can be quickly mass-produced at very high volumes in short periods of time.

Some worry that the virus may mutate faster than the world can immunize everyone for each new variant—not unlike what happens with the annual influenza vaccine process. Dr. Fauci echoed this worry: "My only concern about chasing all the variants is that you'd almost be playing Whac-A-Mole, you know, because they'll keep coming up and keep coming up."[55]

To avoid a perpetual cycle of crises from new Covid variants, some scientists are working on a universal coronavirus vaccine.[56] In theory, a universal vaccine would prime the immune system to react to multiple variants, such as by targeting some part of the virus that does not mutate—probably because it performs some essential function for the virus. Dr. Fauci told the *Atlantic*, "It just makes sense to me to use all of our capabilities to really go for a universal SARS-CoV-2 vaccine. If we don't, we're going to be constantly chasing things, as opposed to getting it off the table."[57] Nobel Laureate and MIT Professor Phillip Sharp went a step further, saying: "In the future, mRNA vaccine technology may allow for one vaccine to target multiple diseases."[58]

Keeping Darwin from Winning

"We don't have evolution on our side," said Devi Sridhar, a professor of public health at the University of Edinburgh.[59] Thinking about the virus through a lens of evolution involves looking at two phenomena: the virus's mutation rate and the various natural selection processes affecting the survival and spread of those mutants. The first phenomenon embodies the opportunities the virus has for mutating. The second one includes the process by which the environment (and human behavior) either suppresses those mutations or enables them to replicate more widely, as a function of natural selection.

The first phenomenon, mutation, is mostly about numbers: The total number of times the virus (and each virus variant) gets to replicate determines the number of potential mutated copies of the virus. The higher the number of infected individuals and the longer the virus is allowed to replicate inside more victims, the greater the chance of producing more mutations.

The second phenomenon, natural selection, is illustrated by the famous quote (wrongly) attributed to Charles Darwin: "'It is not the strongest of the species that survives, nor the most intelligent. It is the one most adaptable to change." Among all the mutations, those that survive are those that are the most infectious; they can find more and more susceptible humans to infect, and they can continue to replicate themselves. Mutations that are not particularly infectious simply disappear due to lack of hosts. Furthermore, because of their higher infection rates, these viral mutations will become the majority of the viruses in circulation. "This pathogen seems to always be changing in a way that makes it harder for us to suppress," Sridhar said.[60] That's because human practices that only partially suppress the spread of the virus actually favor the toughest and most dangerous variants, which are more capable of withstanding such practices, making it harder to snuff out the pandemic.

This aspect of evolution is behind Dr. Fauci's warning at the end of May 2021: "We cannot abandon public health measures when you still have a degree of viral activity in the broad community in the United States. Although we're down to less than 30,000 infections† per day, that's still a lot of infections per day."[61] The more that infected people commingle with potentially immune people, the higher the likelihood of breakthrough infections by mutations that evade the immune system and evolve to infect the vaccinated and the previously infected alike. In other words, uncontrolled spread of Covid provides the perfect petri dish for further evolution and more dangerous strains of the virus. Dr. Maria Van Kerkhove, of the WHO's Health Emergencies Programme, said, "We don't want to be in a situation where the virus changes enough

---

† By the end of August 2021, the number of US infections had catapulted to more than 150,000 per day.

that we go back to square one. This is why we need to prevent as many infections as we can right now."[62]

Although "flattening the curve" was a key strategy early in the pandemic to prevent the collapse of healthcare systems, it created case-rate plateaus with ongoing rates of infection. "Case plateaus can hide the emergence of new variants," said Carl Pearson, a research fellow at the London School of Hygiene & Tropical Medicine.[63] If the effective reproduction number (R_e) of an older, prevalent variant drops below 1 while the effective reproduction number of a newer more dangerous variant rises above 1, the net result might be stable or even declining case rates for a limited period, all the while the dangerous variant is spreading rapidly. That's exactly what happened with the Delta variant in the spring of 2021, which surged in the background.

Each variant has its own reproduction rate and unique proclivities toward symptoms, disease, and death in people of various ages and immunity conditions. "The best way to think about B.1.1.7 [Alpha] and other variants is to treat them as separate epidemics," said Sebastian Funk, a professor of infectious disease dynamics also at the London School of Hygiene & Tropical Medicine. "We're really kind of obscuring the view by adding them all up to give an overall number of cases."[64] This danger highlights the need for systematic, ongoing genetic sequencing of enough samples to catch both new variants and unexpected surges of more virulent ones.

The Race Is On

Thus, humanity is in a deadly race against the evolution of Covid with its growing number of variants, and as of mid-2021, it seemed that the virus was winning. Less than 14 percent of the world's population had been fully vaccinated (and about 27 percent had had at least one dose).[65] The world was vaccinating at a rate of 0.4 doses per 100 people per day, meaning that it would take almost another year (assuming a two-dose regimen) to reach 75 percent worldwide immunity. (Only a little over 2 percent of the world's population may have gained some immunity by natural infection from Covid,[66] but that may be a woefully low estimate.) "This is crunch time," Dr. Peter Hotez, dean of the National School of Tropical Medicine at Baylor College of Medicine, told CNN. "This is going to be our most difficult period right now in terms of seeing who wins out."[67]

Health officials began warning about the rising threat of variants as soon as the data showed the growing danger they posed. For example, the spread of the Alpha variant in early 2021 prompted Dr. Fauci to say, "The way we can counter [Alpha],[68] which is a growing threat in our country, is to do two things: to get as many people vaccinated as quickly and as expeditiously as possible with a vaccine that we know works against this variant. And finally, to implement the public health measures that we talk about all the time."[69]

The rise and spread of variants in countries around the world highlights the risk in overemphasizing a country-specific response. Without draconian border closures that almost no country seems to be able to sustain, the spread of each new variant from its origin to other countries is inevitable. "As long as there is some degree of activity throughout the world, there's always a danger of variants emerging and diminishing somewhat the effectiveness of our vaccines," Fauci told the *Guardian*.[70]

A more hopeful solution to the variants lies in how quickly the mRNA vaccines that were developed for the first variant can be adjusted for other variants. As mentioned above, this vaccine development technology can be adjusted for use against new variants within a relatively short time. Once a new variant is sequenced, it is relatively easy to formulate and manufacture the appropriate vaccine using the same ingredients and processes as the "classic" one. It basically means a quick edit to the genetic code used for the mRNA.[71]

However, fully eradicating Covid calls for more than just developing a booster vaccine. "For a vaccine to do that, what you would need to do is deploy it around the world sufficiently, completely and rapidly so that you wouldn't give the virus time to develop variants fast enough to get around it," explained Professor Roland Kao, chair of veterinary epidemiology and data science at the University of Edinburgh.[72]

Failure to quickly vaccinate a high enough fraction of the world's population increases the chance of additional deadly vaccine-evading variants leading to new pandemics: Covid-21, Covid-22, and so on. In the meantime, public health measures such as contact tracing, social distancing, mask wearing, regulation of super spreader locations, and enhanced ventilation and cleaning can all help limit the rates of infection while the world catches up on vaccinations. The International Monetary Fund (IMF) estimates that the entire world could be vaccinated for a mere $50 billion, a pittance compared to the $16 trillion spent propping up pandemic-pummeled economies.[73]

The Not-So-Great Pandemic Divide

Average vaccination levels and vaccination rates don't tell the full story of the risk of Covid. Although by July 2021 the world may have been vaccinating 0.4 people per 100 per day, India's daily rate was only 0.3 per 100 and Africa's was a meager 0.04 per 100. According to the IMF, "In the absence of urgent actions, many emerging and developing economies may have to wait until the end of 2022 or later to bring the pandemic under control."[74] As of the writing of this book (fall 2021), that gave Covid another 18 to 24 months to further mutate. Considering that Covid was able to double its infectiousness in less than 18 months, allowing the virus that much time for further mutations seems exceedingly dangerous. "That will be too late not just for those countries but also for the world," the IMF warned.

Even within countries that have vaccinated large percentages of their population, such as the US, the statistics don't always reflect those countries' safety. Although the US had fully vaccinated more than 50 percent of the population by the end of July 2021,[75] the coverage was quite uneven, with a mix of leading and lagging states. For example, as of late September 2021, more than two-thirds of adults were fully vaccinated in states like Massachusetts, Connecticut, Maine, and Vermont. In contrast, full vaccination rates hovered near 40 percent in states such as Alabama, Idaho, and West Virginia. Children have been under-vaccinated; only 42 percent of those 12–17 years old had been vaccinated with one or more doses in the US by September—again with wide disparities (for example, 70.1 percent in Vermont and only 20.2 percent in Mississippi).[76]

The Possible Endemic Endgame

The sad truth is that the current vaccines and rates of vaccination may be too low for the world to ever reach herd immunity.[77] The delays in vaccinating everyone in the world,[78] the stubborn prevalence of anti-vaccination sentiments,[79] and the mounting numbers of aggressive mutant strains of Covid[80] may mean that Covid may not disappear for years—or ever.

Instead, many scientists expect Covid to become endemic like influenza[81]—a disease that continues to circulate, with its impact governed by two opposing forces. On one hand, because Covid is both more contagious and more virulent than influenza, ongoing and future outbreaks of Covid variants may sicken and kill significant numbers of people, especially among the unvaccinated and the immunocompromised. On the other hand, as people accumulate exposure to Covid through vaccines and recovering from the virus, they will likely become less prone to illness and death even if they do catch it. The data as of the writing of this book clearly shows that being vaccinated or surviving Covid does provide significant partial immunity against infection by the variants and strong protection against severe sickness or death. Thus, Covid may become much more like the flu: being an unpleasant nuisance for most, posing a danger only to a vulnerable subset of the population, and only occasionally leading to more serious outbreaks for the general public.

Disrupting the Viral Supply Chain

The SARS-CoV-2 virus infection process, like that of any pathogen, depends on a figurative supply chain of fresh victims to infect. The virus depends on "manufacturing" by an infected host to mass-produce more infectious viral particles; "packaging" of the particles in respiratory droplets; "airfreight transportation" via coughing, speaking, singing, and breathing; "arrival at the inbound docks" of enough susceptible victims; and then the hijacking of the manufacturing systems of the next victims. Each one of these steps contributes to the R_e of the virus in terms of the number of subsequent "factories" (i.e., people) hijacked to mass-produce the next batch of viruses.

Although supply chain managers typically do all they can to avoid supply chain disruptions, defeating Covid calls for the opposite goal. Disrupting the supply chain of the virus as much as possible with the least inconvenience and cost to people and society can reduce the spread of infection and, potentially, eliminate the virus. Many possible tactics could disrupt the virus' supply chain and chip away at the reproduction number of the virus, reducing the effective reproduction number (R_e) down from the original R_0 value. Recall that if R_e drops to below 1, the daily infection rate starts shrinking (see The Advantages of Adoption, p. 70).

The first tactic for disrupting the viral supply chain is immunization (e.g., vaccination or recovering from a bout of the disease), which is like training the factory's staff and security guards to look out for and detain the virus. However, immunization of people, like the training of employees, is never 100 percent effective. The first-generation vaccines against Covid seem to have upwards of 95 percent effectiveness against the original Covid variant, but their effectiveness declines over time and is lower with the more aggressive variants. For example, estimates suggest an efficacy against the Delta variant of 72–95 percent for the Moderna vaccine and 42–96 percent for the Pfizer vaccine.[82]

The overall effective immunity level of a community is the percentage of immunized people in the community multiplied by the effectiveness of the immunization processes (e.g., vaccination or recovery). The greater the percentage of immunized people in a community, the lower the rate of growth will be in the infection rate in that community. If half (50 percent) of people are immune, then the R_e will be half the original R_0, making it sufficient to control a variant with an $R_0 < 2$, ensuring that it does not spread. If three-quarters (75 percent) have immunity, the R_e will be one-quarter of the original R_0, making it sufficient to control a variant with an $R_0 < 4$. Note that if the vaccine is only 85 percent effective against a Covid variant, getting to 75 percent immunity requires immunizing 88 percent of the entire population. Suppressing a variant with an R_0 of 7.5 (the estimated R_0 for Delta is between 5 and 9.5[83]) requires vaccinating 100 percent of the population with a vaccine that is at least 87 percent effective against that variant.

A second tactic is to apply means of control inside the "factory" by administering pharmaceuticals to reduce the infection rate inside the body. For example, fenofibrate is a cholesterol drug that was found to markedly inhibit SARS-CoV-2 viral replication in *in vitro* animal cells. The drug not only reduced the number of infected cells, but it also appeared to reduce the amount of viral matter outside of cells.[84] As of this writing, two clinical trials of fenofibrate are already underway in the Hospital of the University of Pennsylvania and the Hebrew University of Jerusalem.[85] If these and other Covid pharmaceutical trials prove successful, each "factory" will become less effective at spreading the virus.

In potentially encouraging news, Merck and Ridgeback Biotherapeutics announced on October 1, 2021 the results of Phase 3 trials of an experimental viral pill for Covid patients that could cut the likelihood of dying or even being hospitalized by 50 percent. The pill, molnupiravir, is designed to introduce errors into the genetic code of the virus. It would be the first oral antiviral medication for Covid-19. Preliminary results showed that only 7.3 percent of patients receiving molnupiravir required hospitalization, with zero deaths; while 14.1 percent of the placebo recipients were hospitalized, and eight died.[86] Merck has licensed the drug to generic manufacturers in order to expedite the drug's availability around the world.[87] Other pharmaceutical companies are also developing oral antiviral Covid medications that may conclude clinical trials in 2021, including Pfizer, and partners Atea Pharmaceuticals and Roche.[88]

A third tactic for disrupting the virus's supply chain is to hire inspectors to check "factories" for Covid and prevent those factories from exposing others by taking them offline. That is, testing everyone and isolating the infected cuts off the supply of viral particles that could infect others. While immunization reduces the percentage of susceptible people, testing reduces the percentage of infectious people. A hypothetical perfect testing program would test everybody every day and immediately isolate those who test positive from the susceptible population.

However, as with immunization, even daily testing is not 100 percent effective. The limitation of testing is that, even if the test is perfectly accurate, anyone who tests positive in that daily test could have exposed others since their previous negative test the day before. In addition, the test results (of high-quality testing procedures) take, even in advanced settings, about 24 hours. On average, then, with a daily test, anyone testing positive would have exposed others for half a day, on the average, before being tested (assuming 24 hours since the last negative test). In addition, these people would expose others for a full day before they receive their positive test result and are isolated from the population. The effective reproduction number created by the day following the positive test and the half day of average exposure before the positive test would depend on the reproduction number in the absence of testing and the average duration of the infectious period.

Researchers estimate that, with Covid, people are infectious for an average of nine days.[89] That implies that, with comprehensive daily testing, the infectious only expose other people for a day and a half, on the average, out of those nine days. Thus, with daily testing, infected people would only transmit the virus 17 percent of the time, on the average, compared to the situation with no testing, and the R_e would be 17 percent of the original R_0. Naturally, if getting the test results takes longer or the test is less than 100 percent accurate, R_e would be higher. Similar logic suggests that twice-weekly testing of everyone in a community would create an R_e that is 31 percent of the original R_0,[‡] while weekly testing would have an R_e of 50 percent of the original R_0. Few institutions, however, are equipped for testing their entire populations with these kinds of frequencies (MIT and other Boston-area universities and private schools were opening their campuses in September 2021 with a weekly testing requirement for all). In general, the US tests only 0.25 percent of its population every day, while Austria tests 3.7 percent, and the UAE 2.8 percent.[90] Despite such low testing rates, these testing regimens can be reasonably effective because they focus on those most likely to be infectious: the symptomatic and those who are found through contact tracing to have been exposed to someone who tested positive.

The last category of supply chain disruptions that can be used to control a pandemic is disruption of the "airfreight transportation system" of the virus—somewhere between the "outbound dock" of the infectious (their noses and mouths) and the inbound side of the susceptible. First, if the infectious wear masks (a common pre-Covid social practice in some Asian societies that could become a more common practice globally), their masks can catch the outbound respiratory droplets from taking to the air. Second, if the susceptible wear masks, their masks will block the inbound droplets from reaching them.[§] Third, social distancing, especially indoors, puts susceptible hosts beyond the typical range of airborne viral particles. (Note, however, that the ubiquitous "6-foot rule" is much less significant than masking, air filtration, and time spent inside.)[91] Fourth, if the indoor places occupied by the infectious and the susceptible have good ventilation, the air that includes virus aerosols is replaced frequently, so these aerosols have less time to be inhaled and infect the susceptible.

Given the limited flight range of viral particles, the ultimate distancing tactic may be more reliance on work from home, online retail, and online service channels (e.g., online education, worship, exercise, telemedicine). The greater the fraction of formerly in-person activities that can be converted into at-home, online, or mobile services, the lower the chances of exposure between the infectious and the susceptible. Although people will still want and need to gather in person, those present at in-person gatherings can be protected by vaccination and

---

[‡] $(3.5 \div 2 + 1) \div 9 = 0.31$

[§] Only the highest quality, best-fitting masks provide protection to the wearer.

testing to reduce the chance of the infectious connecting with the susceptible in close quarters.

Implications for the Public and Businesses

The potential for endemic Covid has two types of implications for people and institutions. The first is in taking steps to minimize the risks of the spread of Covid among staff, customers, and local communities. That implies encouraging (even mandating[92]) vaccination among staff. In fact, many companies, including Google, Facebook, Twitter, Netflix, Lyft, Tyson Foods, Morgan Stanley, Microsoft, Saks Fifth Avenue, The Washington Post, Ascension Health, and BlackRock have announced vaccination mandates, as have many universities, as mentioned above (for students as well as all staff). It is, of course, no surprise that mandates are working and leading to significant increase in vaccinations.[93]

In the same vein, organizations should also create a disease-averse culture, such as setting expectations for those with any sort of illness to either wear masks or work from home. Although these measures may be motivated by Covid, they will also reduce the burden of influenza, colds, pneumonia, and other respiratory infections in the workplace and the wider community.

The second implication of Covid becoming an endemic phenomenon is in preparing the organization for outbreak readiness—enabling companies to easily switch the operating procedures of local offices from "normal" to "outbreak" mode based on local Covid intensities. Retail stores, for example, commonly have different operating procedures for different seasons because they entail different volumes of customers (e.g., extra staff for the holidays) and different types of merchandise (e.g., bulky outdoor gear in summer). Similarly, companies can prepare for "Covid season" and "non-Covid season" either to comply with local rules or as part of a proactive response to potential outbreaks in communities where they operate.

Bioengineering the Future

The near-simultaneous development of more than a dozen success-ful Covid vaccines in less than a year piggybacked on a great many innovations in biomedical technology, such as the understandings of cell biology, genetics, immunology, and virology. Decades of scientific breakthroughs enabled new laboratory, manufacturing, and clinical technologies related to vaccines and diseases. Moreover, these acceler-ated successes actually portend an even brighter future, because they have proven the practicality of some extremely powerful ideas: how to safely program a patient's own cells to effectively fight a disease or improve their health. "It's absolutely astonishing that this has happened in such a short time—to me, it's equivalent to putting a person on the moon," said Dr. Cody Meissner, pediatric infectious-disease specialist at Tufts Children's Hospital in Boston. "This is going to change vaccinol-ogy forever."[94]

The New Lean, Mean Vaccine Machine

Pamela Bjorkman, a bioengineering professor at Caltech, wonders what Covid and its relatives are going to do for an encore. "This isn't going to be the last one," she says of the pandemic. "We're going to have SARS-CoV-3 and SARS-CoV-4. Everyone said this before the current pandemic. Most of the world ignored them. To do so again would really be burying your head in the sand."[95]

Swine flu, bird flu, cowpox, as well as Covid's contorted and con-tested origins[96] (possibly from bats or pangolins) all attest to the dan-gers of zoonosis: diseases transmissible from animals to humans. New diseases, of which 75 percent have come from animals, have literally plagued humanity since the dawn of civilization. Moreover, the spread of human civilization across all parts of the globe, growth of human density with urbanization, and increase in global travel is bringing more people into contact with sick animals and with each other.

What has improved are society's tools to understand the pro-cesses by which invisible viruses, bacteria, or fungi that never afflicted humanity in the past are likely to turn nasty. The rapidly falling costs of genetic testing and sequencing have created a surge in knowledge of the myriad pathogens that afflict animals in the natural world. For example, the United States Agency for International Development (USAID)'s PREDICT pandemic-preparedness project collected half a million samples from 75,000 animals and found 700 new viruses.[97] The project is ranking these and other animal viruses according to their risk of spillover: the chance the pathogen might become much more dangerous to humans.[98] Interestingly, the Covid virus, SARS-CoV-2, ranked right between two viruses that caused multiple outbreaks of hemorrhagic fever in Africa: Lassa and Ebola. Like the other two, the original host of the SARS-CoV-2 was not known before the outbreak and reflects the ongoing risks of zoonosis that can be reduced through programs such as PREDICT. (Likewise, the hosts of SARS—palm civet cats—and MERS—camels—weren't known until these coronaviruses became human diseases.)

"When they write the history of vaccines, this will probably be a turning point," said Amesh Adalja of the Johns Hopkins Center for Health Security, in Maryland.[99] The ability to rapidly sequence a new pathogen, engineer a suitable mRNA sequence, and rapidly mass-produce a vaccine, offers a new platform for tackling pandemics. Andreas Kuhn, senior vice president for RNA biochemistry and manufacturing at BioNTech, said, "One process can be used to manufacture essentially any mRNA sequence."[100] The new technology largely avoids the one-bug-one-drug, trial-and-error processes of attempting to culture the new pathogen in the lab, finding a way to eliminate the pathogen's virulence while retaining its antibody-stimulating properties, and then scaling that pathogen-specific process.

"New advancements in mRNA vaccines and other novel vaccines may mean that we are entering a new golden age for vaccines," said Charles Christy, head of commercial solutions of Ibex Dedicate, the customized supply chain solution developed by Lonza Pharma & Biotech (the contract manufacturer helping mass-produce the Moderna vaccine).[101] Already, Moderna and BioNTech are looking beyond Covid toward developing or testing nine more vaccines each. Targets include influenza, HIV, Nipah, Zika, herpes, dengue, hepatitis, and malaria.[102] "What the success of the coronavirus vaccine has now shown is that mRNA is really a proven technology for infectious diseases," says BioNTech's Kuhn.[103] Similarly, Johnson & Johnson's Janssen unit built what they call AdVac: a viral vector vaccine technology platform they've used to develop an Ebola vaccine, their current Covid vaccine, and three new candidate vaccines (for Zika, RSV, and HIV).

New Messengers for New Missions

Messenger RNA has many potential applications beyond vaccines and other antibody-stimulating products. For example, the acronym VEGF stands for Vascular Endothelial Growth Factor, which is a protein-based hormone that the human body creates to spur the growth of new blood vessels, a process known as angiogenesis. A collaboration between AstraZeneca and Moderna is using mRNA to create VEGF in cells in the hearts of people suffering from heart failure.[104] Early results from clinical trials show that it can trigger regenerative angiogenesis that could restore heart muscle damaged by heart attacks, hypertension, and other heart-related ailments.[105] If the VEGF treatment works for hearts, it may be adaptable to other vascular diseases, such as chronic kidney disease.

Professor Robert Langer of MIT noted the large role of protein-based pharmaceuticals that may be amenable to mRNA approaches: "The last time I looked at the top 10 best-selling drugs, seven of them were proteins, many are antibodies."[106] As a broader technology, mRNA is a way to get the patient's own body to make therapeutic proteins. Messenger RNA is ideal in creating a targeted, short-term boost to the human body's own natural growth and repair mechanisms with less likelihood of side effects compared with small molecule drugs and DNA therapies.

Messenger RNA is but one type of RNA that can be used to program cells. Another key type is called siRNA (small interfering RNA or silencing RNA) that can tell cells to stop making something, such as a key component of a disease-causing pathogen, cancerous tumor, or other cellular product that is harming the patient.[107] While mRNA says, "Make this," siRNA says, "Don't make this."

For decades, scientists have sought ways to cure cancer by getting the patient's own immune system to corral and kill the tumor's proliferating cells. Treating cancer has always been hard because it means creating something toxic enough to kill human cells but not so lethal that it kills the whole human—which is why chemotherapy often makes people so very ill. RNA has two big potential advantages: It can target a very specific cancer-related protein or tumor mutation and then disappear after it performs its duties. MIT's Phillip Sharp noted this advantage for curing cancer, saying, "[there is] the possibility that someone could do that with RNA: arm it, juice it, and put it in the cell, and for another week it would go activate processes to clear out the tumor and then disappear."[108] Treatment would begin with genetic testing or sequencing of the patient's tumor cells to determine the right kind of

RNA therapy. The same production technologies that enabled Moderna to make a candidate vaccine in weeks could let it quickly create a personalized cancer therapy.

The future of these kinds of RNA therapies continues to unfold as scientists further explore the relationship between a cell's DNA and all the cellular activities controlled by it. MIT's Sharp explained, "Keep in mind that all mRNAs in human cells are encoded by only 2 percent of the total genome sequence. Most of the other 98 percent is transcribed into cellular RNAs whose activities remain to be discovered. There could be many future RNA-based therapies."[109]

Engineering Drug Delivery

The use of LNP (lipid nanoparticles) in the vaccines of Pfizer, Moderna, and others to deliver fragile mRNA strings is another category of breakthrough technology with many potential biomedical applications. Currently, many drugs are delivered orally or via injection containing the drug. "That's a big problem," said Pieter Cullis, a University of British Columbia biochemist focusing on LNP, "because the drugs go everywhere in your body, but a very small proportion gets to where you want [it] to go."[110] The result is that much of the drug's dosage is wasted, creating a higher chance of side effects when the drug interacts with other tissues.

The Moderna and Pfizer–BioNTech vaccines both use the same phospholipid in the LNP membrane, but using a different phospholipid could control which cells tend to absorb the LNP, according to Kathryn Whitehead, a nanoparticle scientist at Carnegie Mellon University.[111] Designer LNPs that encapsulate designer mRNA would lead to targeted treatments. In theory, an LNP could be like a shipping box that is addressed to a particular cell type and safely delivers the protected contents of the particle (mRNA or any other therapeutic) directly inside the targeted cell.

"I'm convinced that in the next 50 to 100 years we'll be able to solve all the [medical] problems that we have not yet," said Sylvia Daunert, a biochemist studying nanoparticles at the University of Miami.[112] These nanoparticles, with therapeutic contents and coatings that help direct the particles to their destinations, will be like molecular surgeons traveling through the body to fix particular problems. Technologies such as mRNA will give the body the tools for self-repair. "It's not just The Magic School Bus, it's a reality," Daunert quipped.

Accelerating the Pace of Innovation

Although the Covid mRNA vaccines were the culmination of decades of work, the timeline for their development hints at a future of much faster innovations. When asked about this potential, Pfizer CEO Albert Bourla told the *Washington Post*, "If we did it with Covid, why only Covid? And why can't we speed up treatments for cancer? Why can't we speed up treatments for Alzheimer's or Parkinson's? So, I think that is an experience that will teach us all a lot of things."[113] Daniel Anderson, an mRNA therapy researcher at MIT, remarked, "You can have an idea in the morning, and a vaccine prototype by evening. The speed is amazing."[114]

Sean MacLeod, CEO of Seattle-based FenoLogica Biosciences, a startup combining high-throughput analysis and machine learning to decode how genes affect cells, added, "Recent advances in protein synthesis and genome editing have created the opportunity to rapidly accelerate biotech innovation, transforming the approach to vaccine research, pharmaceutical discovery, and manufacturing."[115] For the most part, scientific knowledge and engineering innovations are cumulative, and medical science and pharmaceutical engineers have accumulated a lot to get to this point.

As these new technologies blossom, though, there is the potential problem that legacy regulatory processes might nip them in the bud. One of the most important challenges will be to reform the regulatory agencies that approve novel treatments. In many cases, current FDA approval processes suppress innovation and delay the application of lifesaving therapeutics.[116] As with other fast-moving technology fields, government cannot keep up. In the high-tech industry, government slowness leads to loss of privacy and the growth of monopolistic enterprises. In biotech, which requires government approvals, the results can be significant delays, lost opportunities, and needless suffering.

Beyond Medicine

Ribosomes are the 3D printers of biological cells. These intricate assemblies of proteins and RNA are tiny programmable machine tools that read an mRNA string of instructions and assemble a corresponding chain of amino acids. The resulting chain of amino acids then naturally folds up into the final product, which may be any of an astounding range of possibilities such as structural proteins, active enzymes, controlling hormones, or other cellular machinery such as more ribosomes.

The mRNA, like an engineering blueprint or a recipe, can be read and translated hundreds or thousands of times to make hundreds or thousands of copies of the product.

The routine application of mRNA and DNA to induce ribosomes to make something useful presents a huge new opportunity for manufacturing a wide range of products that go well beyond medicine. The ribosome could make valuable biomaterials, such as silk for textiles, edible proteins for meat substitutes, or novel bioplastics. Ribosomes can also make enzymes that catalyze valuable chemical reactions, which could make biofuels, digest plastics, neutralize toxic waste, or capture carbon.

Perhaps most importantly, a ribosome can make parts for making more ribosomes. Ribosomes really are like a 3D printer that can print parts for more 3D printers. Self-replication is a key function because it is the pathway to industrial scale. Biological cells are factories, and thus, the most important product of the cell may be more cells.

Whereas a new semiconductor factory can take five years of planning and construction, a single bacterium can make a new bacterium (which is a new bacterium factory) in a mere 20 minutes in the right environment.[117] Although more sophisticated mammalian cells take longer to reproduce, popular biotechnology cell lines such as Chinese hamster ovary cells (CHOs) can double in 20 hours.[118] That implies that CHO cells can increase in factory capacity by 1,000 times every eight days and by a trillion times in a bit more than a month. Then the right added DNA or mRNA can program all those cells to make industrial quantities of anything.

Curing the Climate Pandemic

The challenge of fighting a worldwide pandemic and the challenge of climate change share many common aspects. Both the pandemic and climate change are affecting the entire world, have complex nonlinear effects, and cannot be solved through local initiatives. Global challenges require global solutions.

If the majority of current scientific opinion is correct, climate change could be viewed as a kind of pandemic created by the ongoing, uncontrolled spread of practices that increase the amount of greenhouse gases in the atmosphere (coal-fired power plants, fossil fuel vehicles, deforestation, etc.). The result is a metaphorical infection that affects the entire world as GDPs, standards of living, and consumption of energy and resources rise. This infection is emitting greenhouse gases and other pollutants, likely changing the climate, and damaging the lungs of the planet. The disease threatens to create fevers in some places as higher temperatures increase the frequency of droughts and forest fires. Other places may suffer severe chills; if, for example, a melting Arctic shuts down the Gulf Stream, the climate of Europe may become much more like that of Canada. (In latitude, London is actually north of Winnipeg, Canada, where the average temperature in January is only 3°F [-16°C].)[119]

The Curse of Cultural Inertia

The first line of defense against both climate change and Covid-19 (before vaccines were available) focused on changing behavior and cutting down on certain freedoms and conveniences. To reduce infections during the pandemic, people were asked to wear masks, keep a safe distance from each other, and stop going to bars, restaurants, and other places where people congregate indoors. Much of the current effort to mitigate the future effects of climate change is similarly focused on behavioral changes—doing less: less consumption, less flying, less energy use, and so forth.

Unfortunately, the Covid pandemic revealed how human societies have a great deal of cultural inertia. Even when people's lives were being directly threatened by a potentially fatal disease, a considerable number of people failed to fully comply with recommended behavioral changes. Some people never complied. Others did so for a few months but could not maintain the necessary lifestyle changes to keep the virus at bay. In a sense, the virus was relentless while people relented. More alarming, many people refused to get the lifesaving vaccines even at the risk of severe illness and death.

The threat of climate change is far subtler, far less direct, and farther in the future than the threat of a deadly virus. One of the unfortunate conclusions from the experience of the pandemic is that many people tend to refuse to change their habits even when faced with a daily drumbeat of actual illnesses and deaths. The obvious question is, How can society expect them to change now in response to some nebulous forecast of bad weather some distant decades away? The answer is that many, if not most, people are not going to heed the warning.

Enabling Solutions without Imposing a Change

Given the failure to bring about sufficient behavioral changes to defeat Covid, humanity turned to science and engineering to offer a less onerous solution. In vaccines, the world found a solution that would allow people to continue their normal lives through technology. Scientists around the world used decades of amassed biological and medical knowledge plus collaboration and government funding to develop many highly effective vaccines. Then pharmaceutical organizations built upon global supply chain assets to manufacture and distribute those vaccines, while governments funded and coordinated vaccination campaigns. As outlined throughout this chapter, the first vaccines were only the first technological weapons in the battle to conquer Covid. While manufacturing and distribution operations were ramping up and billions of people were getting vaccinated, the virus mutated, creating more infectious variants. As Covid becomes endemic, the forthcoming technological weapons, which may eradicate it, would be universal vaccines that will work against all variants.

Because most consumers are not yet willing to change their behavior in order to halt climate change, the solution may need to be technological as well. The first phase of developing and using technological solutions to climate change is already here. It includes a range of renewable energy sources. This family of solutions is based on replacing the current carbon-based technologies with clean ones. Accumulating knowledge in physics, chemistry, electronics, engineering, and manufacturing is creating exponential declines in the cost of renewable energy systems.[120] New photovoltaic materials can increase the efficiency and decrease the costs of renewable power; new battery technologies can boost energy density to enable electric-powered transportation[121] and reduce costs of grid storage that can buffer the variable levels of power supplied by solar and wind.[122]

Regardless of their merits, renewable energy sources have inherent limitations. Both wind and solar power, among other limitations, are intermittent and depend on the weather. Both of these renewable technologies require significant land areas, as does biomass energy, among other limitations. However, just like the first generation of vaccines were the first technology salvo in the fight against the pandemic, renewables may be only the first set of technologies to fight climate change. More potent technologies such as carbon capture and sequestration, including air capture, may have to be applied. Such technologies, already operational on a laboratory scale, can not only limit but also reverse the impacts of climate change when developed and applied at scale.

As in the case of the Covid vaccination effort, governments will almost certainly play a key role in funding some of these green technologies on the basis that such technology development and scaling may be far cheaper than the economic and social impacts of climate change. The fact that governments invested billions of dollars in vaccine development and the trillions of dollars to mitigate the economic impacts of the pandemic demonstrates that when governments recognize a danger, money is no object. Such funding is sure to help jump-start R&D into green technologies just as it helped fund the pharmaceutical industry's vaccine development efforts.

Our Tools for the Future

Overall, this book's story of the record-breaking speed and success of the Covid vaccines from lab to jab illustrates important lessons about our world. The story shows the depth and breadth of our societal capacity to invent new technologies, design new products, build production systems, and deliver billions of units of those new products in record time at global scale. Human civilization has created a three-layer foundation—science, engineering, and supply chains—for tackling big global problems.

More Science for More Knowledge

First, scientists have developed a very sophisticated toolbox of scientific methods and data-gathering systems to objectively measure and analyze situations. They've used and refined these systems for decades in an effort to better understand our universe, our world, and ourselves. With these tools and the knowledge gained through them, the 8.8 million scientists of today's world[123] can turn their attention to creating new knowledge to address the world's most pressing problems even beyond public health, including challenges such as poverty, food and water security, and climate change. It is that growing toolbox, knowledge base, and capacity to answer questions that then feeds the next layer of development.

More Engineering for More Know-How

Second, engineers have applied the tremendous troves of accumulated scientific knowledge to develop solutions to problems. New or updated scientific ideas give rise to new engineering formulas, inventions, patents, and know-how. The engineers use their know-how to convert laboratory findings into workshop prototypes and then into designs for factory systems. In turn, this layer of engineering ideas, tools, and designs underpins a vast global array of manufacturing and service organizations.

More Supply Chains for More Products and Services

Third, manufacturing and service organizations have applied engineering know-how to build and operate manufacturing, logistics, and service operations that create and deliver large volumes of specific materials, parts, products, and services. These organizations have built relationships with other organizations to define globe-spanning supply chains that convert the bounty of the planet into all the goods and services used by its 7.8 billion consumers. New scientific ideas and engineering know-how create new opportunities for improved downstream products and services while new perceived problems or needs percolate upstream to drive more engineering and science. The result is that no matter how specialized the product (e.g., enzymes for producing mRNA), chances are that someone somewhere knows how to make it, or has the on-site engineering and scientific staff to create a way to make it.

More Connectivity for More Speed and Effectiveness

These three foundational layers are woven together by ever-expanding global networks of cross connections created by the internet and transportation systems. Mobile broadband, online collaboration apps, video calls, messaging systems, and cloud services enable interdisciplinary global teams both in and across all three layers. Automated data collection, digital networks, big data repositories, and analytics software provide visibility, control, and objective data for optimized management. Even with pandemic disruptions, physical transportation systems continue to move goods (and people) in high volumes and rapid speeds around the globe. The overall result is that scientific knowledge, engineering solutions, and manufacturing capacity are readily available to almost anyone anywhere.

We Are the Limits to Our Own Futures

Alas, the epic success story of the Covid vaccines also revealed a dark underbelly to human civilization: Some people create and spread erroneous information either through misunderstanding, wishful (or fearful) thinking, or for fraudulent gain. Such spreaders of falsehood find receptive ears in times of stress. However, in a powerful knowledge-driven economy, the use of misinformation can create a serious risk of squandered resources and tragic outcomes (e.g., the high death tolls and economic losses in anti-vax strongholds). Although different people can certainly have different values, these individual beliefs and values don't change the actual biological or physical properties of viruses, vaccines, masks, or planetary atmospheres.

The Way Forward

The challenge facing complex pluralistic societies is to imbue the three layers of science, engineering, and supply chains with a robustness against falsehood but also a sensitivity to personal values. Three improvements to knowledge supply chains can help address this sociological problem. The first is to build trust in the institutions that find or create knowledge. That, in turn, requires ensuring that said institutions are, in fact, worthy of that trust among people of all political outlooks. The second improvement is to sufficiently (but lightly) regulate information creation and distribution systems (especially social media) to curtail the spread of misinformation without censoring legitimate differences in personal preferences or values. The third is to educate people so that they can make informed decisions—as consumers and as voters—that are consistent with both their personal, subjective values and objective reality. Although to err is human, making progress on these challenges can mitigate the effects of human errors, regardless of whether they are accidental or intentional.

As the successful invention and launch of the Covid vaccines show, the combined power of science, engineering, and supply chains has created a growing global capacity to confront the unknown, understand it, and improve the future. Moreover, each problem that we tackle (e.g., Covid) creates new assets in the form of new science, engineering, and supply chains, which accelerate solving the next set of critical global challenges.

References

References for Chapter 1

1.   International Federation of Pharmaceutical Manufacturers & Associations. "The Complex Journey of a Vaccine: The Steps Behind Developing a New Vaccine." Geneva, Switzerland: IFPMA. Accessed September 28, 2021. https://www.ifpma.org/wp-content/uploads/2019/07/IFPMA-ComplexJourney-2019_FINAL.pdf

2.   Pronker, Esther S., Tamar C. Weenen, Harry Commandeur, Eric H.J.H.M. Claassen, and Albertus D.M.E. Osterhaus. "Risk in Vaccine Research and Development Quantified." *PLoS One* 8, no. 3 (2013): 357755. doi: 10.1371/journal.pone.0057755

3.   Roos, Dave. "How a New Vaccine Was Developed in Record Time in the 1960s." History.com. June 22, 2020. https://www.history.com/news/mumps-vaccine-world-war-ii

4.   Markel, Howard. "Science Diction: The Origin of the Word 'Vaccine'." Science Friday. November 2, 2015. https://www.sciencefriday.com/articles/the-origin-of-the-word-vaccine/

5.   BioExplorer.net. "History of Immunology." Bio Explorer, September 28 2021. https://www.bioexplorer.net/history_of_biology/immunology/

6.   Pickrell, John. "Timeline: Genetics." *New Scientist* [online]. September 4, 2006. https://www.newscientist.com/article/dn9966-timeline-genetics/

7.   LunaDNA. "The History of DNA." LunaDNA, April 24, 2019. Last edited September 2019. https://www.lunadna.com/blog/history-of-dna/

8.   Matthew Cobb, "Who discovered messenger RNA?" *Current Biology*, 25 no. 13 (2015): R526–32. doi: 10.1016/j.cub.2015.05.032

9.   "3 Questions: Phillip Sharp on the Discoveries that Enabled mRNA Vaccines for Covid-19." *MIT News*. December 11, 2020. https://news.mit.edu/2020/phillip-sharp-rna-vaccines-1211

10. NIH National Human Genome Research Institute. "Genetic Code." Accessed September 28, 2021. https://www.genome.gov/genetics-glossary/Genetic-Code

11.   Dr. Katalin Karikó, Senior Vice President at BioNTech RNA
      Pharmaceuticals, in interview with the author, March 26, 2021.

12.   Johnson, Carolyn Y. "A Gamble Pays Off in 'Spectacular Success':
      How the Leading Coronavirus Vaccines Made It to the Finish Line."
      Washington Post, December 6, 2020. https://www.washingtonpost.com/
      health/2020/12/06/covid-vaccine-messenger-rna/

13.   Ibid.

14.   Dr. Katalin Karikó, Senior Vice President at BioNTech RNA
      Pharmaceuticals, in interview with the author, March 26, 2021.

15.   Johnson, "A Gamble Pays Off in 'Spectacular Success': How the Leading
      Coronavirus Vaccines Made It to the Finish Line."

16.   Cross, Ryan. "Without These Lipid Shells, There Would Be no
      mRNA Vaccines for COVID-19." *Chemical & Engineering News,*
      March 6, 2021. https://cen.acs.org/pharmaceuticals/drug-delivery/
      Without-lipid-shells-mRNA-vaccines/99/i8

17.   Phillip Sharp, Nobel Prize-winning biochemist and Institute Professor at
      MIT, in interview with the author, March 17, 2021.

18.   Cross, "Without These Lipid Shells, There Would Be no mRNA Vaccines
      for COVID-19."

19.   The University of British Columbia Life Sciences Institute. "LSI Spinoff
      Company the Source of Technology Pfizer's mRNA COVID-19 Vaccine
      'Can't Work Without'." November 13, 2020. https://lsi.ubc.ca/2020/11/13/
      lsi-spinoff-company-the-source-of-technology-Pfizers-mrna-covid-19-
      vaccine-cant-work-without/

20.   Herper, Matthew, Damian Garde, and Helen Branswell. "Studies Provide
      Glimpse at Efficacy of Covid-19 Vaccines from Oxford-AstraZeneca
      and CanSino." *Stat News,* July 20,2020. https://www.statnews.
      com/2020/07/20/study-provides-first-glimpse-of-efficacy-of-oxford-
      AstraZeneca-Covid-19-vaccine/

21.   Johnson, "A Gamble Pays Off in 'Spectacular Success': How the Leading
      Coronavirus Vaccines Made It to the Finish Line."

22.   Ustinova, Anastasia. "In the Thick of the 'Herculean' Vaccine Push."
      SME Media, September 21, 2020. https://www.sme.org/technologies/
      articles/2020/september/vaccine-placeholder/

23.   Centers for Disease Control and Prevention. "CDC SARS Response
      Timeline." Page last reviewed: April 26, 2013. Accessed September 28,
      2021. https://www.cdc.gov/about/history/sars/timeline.htm

24. Milne-Price, Shauna, Kerri L. Miazgowicz, and Vincent J. Munster. "The emergence of the Middle East Respiratory Syndrome coronavirus." *Pathogens and Disease* 71 no. 2 (2014): 121–36. doi: 10.1111/2049-632X.12166

25. World Health Organization. "Middle East Respiratory Syndrome Coronavirus (MERS-CoV)." Accessed September 28, 2021. https://www.who.int/health-topics/middle-east-respiratory-syndrome-coronavirus-mersl

26. Hixenbaugh, Mike. "Scientists were close to a coronavirus vaccine years ago. Then the money dried up." *NBC News,* March 5, 2020. Updated March 8, 2020. https://www.nbcnews.com/health/health-care/scientists-were-close-coronavirus-vaccine-years-ago-then-money-dried-n1150091

27. Holmes, Edward C., on behalf of the consortium led by Professor Yong-Zhen Zhang, Fudan University, Shanghai.. "Novel 2019 Coronavirus Genome." [listserv notice]. January 10, 2020. https://virological.org/t/novel-2019-coronavirus-genome/319

28. "Coronavirus Timeline." The History of Vaccines. Accessed September 28, 2021. https://www.historyofvaccines.org/content/articles/coronavirustimeline

29. The Semantic Scholar [website]. Accessed September 28, 2021. https://www.semanticscholar.org

30. "Get Started with CORD-19" Semantic Scholar. Accessed September 28, 2021. https://www.semanticscholar.org/cord19/get-started

31. Ustinova, "In the Thick of the 'Herculean' Vaccine Push."

32. Irwin, Neil. "The Pandemic Is Showing Us How Capitalism Is Amazing, and Inadequate." *New York Times,* November 14, 2020. https://nyti.ms/38GEPOO

33. Wallace-Wells, David. "We Had the Vaccine the Whole Time." *New York Magazine,* December 7, 2020. https://nymag.com/intelligencer/2020/12/Moderna-Covid-19-vaccine-design.html

34. Loftus, Peter. "Drugmaker Moderna Delivers First Experimental Coronavirus Vaccine for Human Testing", *Wall Street Journal,* February 24, 2020. https://www.wsj.com/articles/drugmaker-Moderna-delivers-first-coronavirus-vaccine-for-human-testing-11582579099

35. Markel, "Science Diction: The Origin of the Word 'Vaccine'."

36. "Vaccine Development, Testing, and Regulation." The History of Vaccines. Last Update January 17, 2018. Accessed September

28, 2021. https://www.historyofvaccines.org/content/articles/
vaccine-development-testing-and-regulation

37.   Muñoz-Fontela, César, William E. Dowling, Simon G. P. Funnell, et al.
      Animal Models for COVID-19. *Nature* 586 (2020): 509–515. https://www.
      nature.com/articles/s41586-020-2787-6

38.   Sumon, Tofael Ahmed, Md. Ashraf Hussain, Md. Tawheed Hasan,
      et al. "A Revisit to the Research Updates of Drugs, Vaccines, and
      Bioinformatics Approaches in Combating COVID-19 Pandemic." *Frontiers
      in Molecular Biosciences*, January 25, 2021. https://www.frontiersin.org/
      articles/10.3389/fmolb.2020.585899/full

39.   U.S. Food & Drug Administration. "Investigational New
      Drug Applications (INDs) for CBER-Regulated Products."
      Accessed September 28, 2021. https://www.fda.gov/vaccines-
      blood-biologics/development-approval-process-cber/
      investigational-new-drug-ind-or-device-exemption-ide-process-cber

40.   "Vaccine Development, Testing, and Regulation." The History of
      Vaccines.

41.   Zhang, Sarah. "A Vaccine Reality Check." *Atlantic*, July 24, 2020. Updated
      July 25, 2020. https://www.theatlantic.com/health/archive/2020/07/
      Covid-19-vaccine-reality-check/614566/

42.   Le, Tung Thanh, Zacharias Andreadakis, Arun Kumar, et al. "The
      COVID-19 Vaccine Development Landscape." *Nature Reviews
      Drug Discovery*, April 9, 2020. https://www.nature.com/articles/
      d41573-020-00073-5

43.   "Safety and Immunogenicity Study of 2019-nCoV Vaccine (mRNA-1273)
      for Prophylaxis of SARS-CoV-2 Infection (COVID-19)." U.S. National
      Library of Medicine. February 25, 2020. Last updated September 16,
      2021. https://clinicaltrials.gov/show/NCT04283461

44.   Moderna, Inc. "Moderna Completes Enrollment of Phase 2 Study of its
      mRNA Vaccine Against COVID-19 (mRNA-1273)." [Press release] July 8,
      2020. https://investors.Modernatx.com/news-releases/news-release-
      details/Moderna-completes-enrollment-phase-2-study-its-mrna-vaccine

45.   Jamrozik, Euzebiusz, and Michael J. Selgelid. "COVID-19 Human
      Challenge Studies: Ethical Issues." *Lancet Infectious Diseases* 20 no. 8
      (2020): e198–e203. doi: 10.1016/S1473-3099(20)30438-2 https://www.
      ncbi.nlm.nih.gov/pmc/articles/PMC7259898/

46.   McGinty, Jo Craven. "Unlike Clinical Trials, New Covid-19 Study Needed
      Only a Few Volunteers." *The Wall Street Journal*, August 27, 2021. https://

www.wsj.com/articles/unlike-clinical-trials-new-Covid-19-study-needed-only-a-few-volunteers-11630056600?mod=searchresults_pos1&page=1

47.   "Want to join a coronavirus vaccine trial? Consider these 5 things first." Advisory Board. August 7, 2020. https://www.advisory.com/en/daily-briefing/2020/08/07/vaccine-volunteer

48.   Armstrong, Drew. "The World's Most Loathed Industry Gave Us a Vaccine in Record Time." *Bloomberg Businessweek*, December 23, 2020. https://www.bloomberg.com/news/features/2020-12-23/covid-vaccine-how-big-pharma-saved-the-world-in-2020

49.   Grady, Denise. "Moderna and Pfizer Begin Late-Stage Vaccine Trials." *New York Times*, July 27, 2020. https://www.nytimes.com/2020/07/27/health/Moderna-vaccine-covid.html

50.   "A Study to Evaluate Efficacy, Safety, and Immunogenicity of mRNA-1273 Vaccine in Adults Aged 18 Years and Older to Prevent COVID-19." U.S. National Library of Medicine. July 14, 2020; Last updated June 10, 2021. https://clinicaltrials.gov/ct2/show/NCT04470427

51.   Jaklevic, Mary Chris. "Researchers Strive to Recruit Hard-Hit Minorities Into COVID-19 Vaccine Trials." *JAMA Network*, August 13, 2020. https://jamanetwork.com/journals/jama/fullarticle/2769611

52.   Rubin, Rita. "Pregnant People's Paradox—Excluded From Vaccine Trials Despite Having a Higher Risk of COVID-19 Complications." *JAMA Network*, February 24, 2021. https://jamanetwork.com/journals/jama/fullarticle/2777024

53.   "A Study to Evaluate Efficacy, Safety, and Immunogenicity of mRNA-1273 Vaccine in Adults Aged 18 Years and Older to Prevent COVID-19."

54.   "Study to Evaluate the Safety, Tolerability, and Immunogenicity of SARS CoV-2 RNA Vaccine Candidate (BNT162b2) Against COVID-19 in Healthy Pregnant Women 18 Years of Age and Older." U.S. National Library of Medicine. February 15, 2021; Last updated September 17, 2021. https://clinicaltrials.gov/ct2/show/NCT04754594

55.   Stenson, Jacqueline. "The FDA's cutoff for Covid-19 vaccine effectiveness is 50 percent. What does that mean?" *NBC News*, November 3, 2020. https://www.nbcnews.com/health/health-news/fda-s-cutoff-Covid-19-vaccine-effectiveness-50-percent-what-n1245506

56.   Ibid.

57. Pfizer, Inc. "Pfizer and BioNTech Announce Publication of Results from Landmark Phase 3 Trial of BNT162B2 COVID-19 Vaccine Candidate in the New England Journal of Medicine." [Press release] December 10, 2020. https://www.Pfizer.com/news/press-release/press-release-detail/Pfizer-and-BioNTech-announce-publication-results-landmark

58. Buranyi, Stephen. "The mRNA Vaccine Revolution Is Just Beginning." *Wired*, March 6, 2021. https://www.wired.co.uk/article/mrna-vaccine-revolution-katalin-kariko

59. U.S. Food & Drug Administration. "FDA Takes Additional Action in Fight Against COVID-19 By Issuing Emergency Use Authorization for Second COVID-19 Vaccine." [News release] December 18, 2020. https://www.fda.gov/news-events/press-announcements/fda-takes-additional-action-fight-against-Covid-19-issuing-emergency-use-authorization-second-covid [archived]

60. Burki, Talha Khan. "The Russian Vaccine for COVID-19." *Lancet Respiratory Medicine,* September 4, 2020. https://www.thelancet.com/journals/lanres/article/PIIS2213-2600(20)30402-1/fulltext

61. Zimmer, Carl, Jonathan Corum, and Sui-Lee Wee. "Coronavirus Vaccine Tracker." *New York Times.* Updated September 28, 2021. https://nyti.ms/2MHNdRL

62. Merck. "Merck Discontinues Development of SARS-CoV-2/COVID-19 Vaccine Candidates; Continues Development of Two Investigational Therapeutic Candidates." [News release] January 25, 2021. https://www.merck.com/news/merck-discontinues-development-of-SARS-CoV-2-Covid-19-vaccine-candidates-continues-development-of-two-investigational-therapeutic-candidates/

63. AstraZeneca. "AZD1222 US Phase III Primary Analysis Confirms Safety and Efficacy." [News release] March 25, 2021. https://www.AstraZeneca.com/content/astraz/media-centre/press-releases/2021/azd1222-us-phase-iii-primary-analysis-confirms-safety-and-efficacy.html

64. Herper, Garde, and Branswell, "Studies Provide Glimpse at Efficacy of Covid-19 Vaccines from Oxford-AstraZeneca and CanSino."

65. Covid-19 Vaccine Tracker [website]. Accessed August 9, 2021. https://Covid-19.trackvaccines.org/vaccines/

66. Ibid.

67. AJMC Staff. "A Timeline of COVID-19 Developments in 2020." *American Journal of Managed Care*, January 1, 2021. https://www.ajmc.com/view/a-timeline-of-Covid-19-developments-in-2020; European

Medicines Agency. "EMA Recommends First COVID-19 Vaccine for Authorisation in the EU." [News release] December 21, 2020. https://www.ema.europa.eu/en/news/ema-recommends-first-Covid-19-vaccine-authorisation-eu; European Medicines Agency. "EMA Recommends COVID-19 Vaccine Moderna for Authorisation in the EU." [News release] June 1, 2021. https://www.ema.europa.eu/en/news/ema-recommends-Covid-19-vaccine-Moderna-authorisation-eu

68.   Cohen, Sandy. "The Fastest Vaccine in History." UCLA Health. December 10, 2020. https://connect.uclahealth.org/2020/12/10/the-fastest-vaccine-in-history/

69.   Armstrong, "The World's Most Loathed Industry Gave Us a Vaccine in Record Time."

70.   U.S. Department of Health and Human Services. "Trump Administration Announces Framework and Leadership for 'Operation Warp Speed'." [Press release] May 15, 2020. https://www.hhs.gov/about/news/2020/05/15/trump-administration-announces-framework-and-leadership-for-operation-warp-speed.html

71.   Higgins-Dunn, Noah. "The U.S. has already invested billions in potential coronavirus vaccines. Here's where the deals stand." *CNBC*, August 14, 2020. https://www.cnbc.com/2020/08/14/the-us-has-already-invested-billions-on-potential-coronavirus-vaccines-heres-where-the-deals-stand.html

72.   Steenhuysen, Julie, Peter Eisler, Allison Martell, Stephanie Nebehay. "Countries, companies risk billions in race for coronavirus vaccine." Reuters, April 27, 2020. https://www.reuters.com/article/health-coronavirus-vaccine/special-report-countries-companies-risk-billions-in-race-for-coronavirus-vaccine-idUSL2N2CF0JG

73.   World Health Organization. "COVAX." Accessed September 28, 2021. https://www.who.int/initiatives/act-accelerator/covax

74.   Winsor, Morgan. "What is COVAX? How a global initiative is helping get COVID-19 vaccines to poorer countries." *ABC News*, February 26, 2021. https://abcnews.go.com/Health/covax-global-initiative-helping-Covid-19-vaccines-poorer/story?id=76106981

75.   "G7 countries commit $7.5B to Covax vaccine funding - latest updates." TRT World. February 19, 2021. https://www.trtworld.com/life/g7-countries-commit-7-5b-to-covax-vaccine-funding-latest-updates-44304

76. Bill & Melinda Gates Foundation. "Bill and Melinda Gates Pledge $10 Billion in Call for Decade of Vaccines." [Press release]. Accessed September 28, 2021. https://www.gatesfoundation.org/Ideas/Media-Center/Press-Releases/2010/01/Bill-and-Melinda-Gates-Pledge-$10-Billion-in-Call-for-Decade-of-Vaccines

77. Peters, Adele. "Gates Versus the Pandemic." *Fast Company*. Accessed September 28, 2021. https://www.fastcompany.com/90579390/inside-the-gates-foundations-epic-fight-against-Covid-19

78. Cheney, Catherine. "Gates Foundation COVID-19 commitment reaches $1.75B with latest pledge." Devex, December 10, 2020. https://www.devex.com/news/gates-foundation-Covid-19-commitment-reaches-1-75b-with-latest-pledge-98739

79. Brennan, Margaret. "Transcript: Pfizer CEO Dr. Albert Bourla on "Face the Nation," September 13, 2020." *CBS News*, September 13, 2020. https://www.cbsnews.com/news/transcript-Pfizer-ceo-dr-albert-bourla-on-face-the-nation-september-13-2020/

80. Hopkins, Jared S. "How Pfizer Delivered a Covid Vaccine in Record Time: Crazy Deadlines, a Pushy CEO." *Wall Street Journal*, December 11, 2020. https://www.wsj.com/articles/how-Pfizer-delivered-a-covid-vaccine-in-record-time-crazy-deadlines-a-pushy-ceo-11607740483

81. Ottesen, K.K. "Pfizer CEO on the pressures of creating a covid-19 vaccine: 'What is at stake is beyond imagination'." *Washington Post*, September 29, 2020. https://www.washingtonpost.com/lifestyle/magazine/Pfizer-ceo-on-the-pressures-of-creating-a-Covid-19-vaccine-what-is-at-stake-is-beyond-imagination/2020/09/29/2ddbe7f8-fdb3-11ea-b555-4d-71a9254f4b_story.html

82. "Pfizer's COVID-19 vaccine is first to win U.S. approval." *Fortune*, December 11, 2020. https://fortune.com/2020/12/11/Covid-19-vaccine-Pfizer-fda-approval/

83. European Medicines Agency. "EMA Recommends First COVID-19 Vaccine for Authorisation in the EU."

84. Covid-19 Vaccine Tracker [website]. Accessed July 28, 2021. https://Covid-19.trackvaccines.org/vaccines/

85. Mishra, Manas, and Michael Erman. "Pfizer says 2021 COVID-19 vaccine sales to top $33.5 bln, sees need for boosters." Reuters, July 28, 2021. https://www.reuters.com/business/healthcare-pharmaceuticals/Pfizer-raises-estimates-2021-sales-Covid-19-vaccine-335-bln-2021-07-28/

86.   Gates, Bill. "Here's How to Make Up for Lost Time on Covid-19."
       Washington Post, March 31, 2020. https://www.washingtonpost.
       com/opinions/bill-gates-heres-how-to-make-up-for-lost-time-on-
       covid-19/2020/03/31/ab5c3cf2-738c-11ea-85cb-8670579b863d_story.
       html

87.   U.S. Department of Defense. "Trump Administration Collaborates
       With McKesson for COVID-19 Vaccine Distribution." [News release]
       August 14, 2020. https://www.defense.gov/News/Releases/Release/
       Article/2313808/trump-administration-collaborates-with-mckesson-for-
       covid-19-vaccine-distributi/

88.   Keith, Tamara. "How The White House Got 2 Pharma Rivals To Work
       Together On COVID-19 Vaccine." *NPR*, March 3, 2021. https://www.npr.
       org/2021/03/03/973117712/how-the-white-house-got-2-pharma-foes-to-
       work-together-on-Covid-19-vaccine

References for Chapter 2

1.   Author interview with Meri Stevens, Worldwide Vice President Supply Chain – Consumer Health and Deliver at Johnson & Johnson, MIT CTL webinar/interview, March 17, 2021

2.   Nordqvist, Joseph. "Health Benefits and Risks of Myrrh." *Medical News Today,* May 21, 2018. https://www.medicalnewstoday.com/articles/267107

3.   Van Arnum, Patricia. "API Sourcing: The Supply Side for US-Marketed Drugs." DCAT Value Chain Insights. November 20, 2019. https://www.dcatvci.org/6213-global-api-sourcing-which-countries-lead

4.   Rägo, Lembit, and Budiono Santoso. "Drug Regulation: History, Present and Future." In *Drug Benefits and Risks: International Textbook of Clinical Pharmacology,* revised 2nd ed., edited by Chris J. van Boxtel, Budiono Santoso, and I. Ralph Edwards. Amsterdam and Uppsala, Sweden: IOS Press and The Uppsala Monitoring Centre, 2008.

5.   International Society for Pharmaceutical Engineering. "What Is GMP?" Accessed September 30, 2021. https://ispe.org/initiatives/regulatory-resources/gmp/what-is-gmp

6.   McGee Pharma International. "EU and US GMP/GDP: Similarities and Differences." November 2016. Accessed September 29, 2021. https://www.pda.org/docs/default-source/website-document-library/chapters/presentations/new-england/eu-and-us-gmp-gdp-similarities-and-differences.pdf?sfvrsn=8

7.   Owermohle, Sarah. "The 'Biggest Challenge' Won't Come Until After a Coronavirus Vaccine Is Found." *Politico,* May 11, 2020. https://www.politico.com/news/2020/05/11/coronavirus-vaccine-supply-shortages-245450

8.   Paris, Costas. "Supply-Chain Obstacles Led to Last Month's Cut to Pfizer's Covid-19 Vaccine-Rollout Target." *Wall Street Journal,* December 3, 2020. https://www.wsj.com/articles/Pfizer-slashed-its-Covid-19-vaccine-rollout-target-after-facing-supply-chain-obstacles-11607027787

9.   Sheffi, Yossi. *The New (Ab)Normal: Reshaping Business and Supply Chain Strategy Beyond Covid-19.* Cambridge, Mass.: MIT CTL Media, 2020.

10. Woolston, Chris. "'Does Anyone Have Any of These?' Lab-Supply Shortages Strike Amid Global Pandemic." *Nature,* March 9, 2021. https://www.nature.com/articles/d41586-021-00613-y

11. Ibid.

12.   U.S. Food & Drug Administration. "Medical Device Shortages During the COVID-19 Public Health Emergency." Last updated September 10, 2021. Accessed September 29, 2021. https://www.fda.gov/medical-devices/coronavirus-Covid-19-and-medical-devices/medical-device-shortages-during-Covid-19-public-health-emergency#shortage

13.   Zhang, Sarah. "America Is Running Low on a Crucial Resource for COVID-19 Vaccines." *Atlantic*, August 31, 2020. https://www.theatlantic.com/science/archive/2020/08/america-facing-monkey-shortage/615799/

14.   Vanderklippe, Nathan. "Chinese Wildlife Ban Freezes Export of Test Monkeys Amid Worldwide Push for COVID-19 Vaccine." *Globe and Mail*, April 2, 2020. https://www.theglobeandmail.com/world/article-chinese-wildlife-ban-freezes-export-of-test-monkeys-amid-worldwide/

15.   Wee, Sui-Lee. "Future Vaccines Depend on Test Subjects in Short Supply: Monkeys." *New York Times*, February 23, 2021. https://www.nytimes.com/2021/02/23/business/covid-vaccine-monkeys.html

16.   Feister, Alan J., Anna DiPietrantonio, Jeffrey Yuenger, Karen Ireland, Anjana Rao. "Nonhuman Primate Evalutation and Analysis. Part 1: Analysis of Future Demand and Supply." National Institutes of Health Office of Research Infrastructure Programs. September 21, 2018. https://orip.nih.gov/sites/default/files/508%20NHP%20Evaluation%20and%20Analysis%20Final%20Report%20-%20Part%201%20Update%2030Oct2018_508.pdf

17.   Born, Richard T. "Unfriendly Skies for Medical Innovation." *Wall Street Journal*, November 17, 2019. https://www.wsj.com/articles/unfriendly-skies-for-medical-innovation-11574018639

18.   Wee, "Future Vaccines Depend on Test Subjects in Short Supply: Monkeys."

19.   Zhang, Sarah "America Is Running Low on a Crucial Resource for COVID-19 Vaccines." The Atlantic, August 21, 2020. https://www.theatlantic.com/science/archive/2020/08/america-facing-monkey-shortage/615799/

20.   British Society for Immunology. "A Chicken's Egg (1931)." Accessed September 29, 2021. https://www.immunology.org/chickens-egg-1931

21.   Boerner, Krietsch. "The Flu Shot and the Egg." *ACS Central Science* 6 no. 2 (2020): 89–92. https://www.ncbi.nlm.nih.gov/pmc/articles/PMC7047267/

22. Ustinova, Anastasia. "In the Thick of the 'Herculean' Vaccine Push." SME Media, September 21, 2020. https://www.sme.org/technologies/articles/2020/september/vaccine-placeholder/

23. Stanton, Dan. "Moderna Says 'Simple' mRNA Process Allowed Speedy COVID Vaccine Scale-Up." BioProcess International, September 23, 2020. https://bioprocessintl.com/bioprocess-insider/facilities-capacity/Moderna-says-simple-mrna-process-allowed-speedy-covid-vaccine-scale-up/

24. Author interview with Paul Granadillo, Senior Vice President of Supply Chain at Moderna, April 22, 2021.

25. Ibid.

26. ModernaTX, Inc. "A Phase 3, Randomized, Stratified, Observer-Blind, Placebo-Controlled Study to Evaluate the Efficacy, Safety, and Immunogenicity of mRNA-1273 SARS-CoV-2 Vaccine in Adults Aged 18 Years and Older." December 23, 2020. https://www.Modernatx.com/sites/default/files/content_documents/Final%20mRNA-1273-P301%20Protocol%20Amendment%206%20-%23Dec2020.pdf

27. McCoy, Michael. "Lipids, the Unsung COVID-19 Vaccine Component, Get Investment." *Chemical & Engineering News*, February 12, 2021. http://cen.acs.org/business/outsourcing/Lipids-unsung-Covid-19-vaccine/99/web/2021/02?PageSpeed=noscript

28. "Moderna Extends Lipid Supply for Coronavirus Vaccine" [Press release]. CordenPharma, May 28, 2020. https://www.CordenPharma.com/CordenPharma_and_Moderna_extend_Lipid_Supply_Agreement_for_Moderna_Vaccine_mRNA-1273_Against_Novel_Coronavirus_SARS-CoV-2

29. Mullin, Rick. "Pfizer, Moderna Ready Vaccine Manufacturing Networks." *Chemical & Engineering News*, November 25, 2020. https://cen.acs.org/business/outsourcing/Pfizer-Moderna-ready-vaccine-manufacturing/98/i46

30. Heilweil, Rebecca. "The Key Ingredient That Could Hold Back Vaccine Manufacturing." *Vox*, March 3, 2021. https://www.vox.com/22311268/covid-vaccine-shortage-Moderna-Pfizer-lipid-nanoparticles

31. Elveflow, "Microfluidics: A general review of microfluidics." February 5, 2021. https://www.elveflow.com/microfluidic-reviews/general-microfluidics/a-general-overview-of-microfluidics/

32. Delaquilla, Alessandra. "History of Microfluidics." Elveflow,
     February 5, 2021. https://www.elveflow.com/microfluidic-reviews/
     general-microfluidics/history-of-microfluidics/

33. Rowland, Christopher. "Inside Pfizer's race to produce
     the world's biggest supply of covid vaccine." *Washington
     Post,* June 16, 2021. https://www.washingtonpost.com/
     business/2021/06/16/Pfizer-vaccine-engineers-supply/

34. Cott, Emma, Elliot deBruyn, and Jonathan Corum. "How Pfizer Makes Its
     Covid-19 Vaccine," *New York Times*, April 28, 2021, https://www.nytimes.
     com/interactive/2021/health/Pfizer-coronavirus-vaccine.html

35. Rowland, "Inside Pfizer's race to produce the world's biggest supply of
     covid vaccine."

36. Author interview with Paul Granadillo, Senior Vice President of Supply
     Chain at Moderna, April 22, 2021.

37. Hantzinikolas, Jenny. "Single Use Technology: A Regulatory Perspective."
     [Slide presentation] Australian Government Department of Health,
     Therapeutic Goods Administration. November 15, 2016. https://www.
     tga.gov.au/sites/default/files/tga-presentation-ispe-meeting-15-
     nov-2016-161124.pdf

38. Sinclair, Andrew, Lindsay Leveen, Miriam Monge, Janice Lim, and
     Stacey Cox. "The Environmental Impact of Disposable Technologies."
     BioPharm International, November 2, 2008. https://web.archive.
     org/web/20110711002623/http://biopharminternational.findpharma.
     com/biopharm/article/articleDetail.jsp?id=566014

39. Rowland, "Inside Pfizer's race to produce the world's biggest supply of
     covid vaccine."

40. Rowland, "Inside Pfizer's race to produce the world's biggest supply of
     covid vaccine."

41. Sealy, Amanda. "Manufacturing Moonshot: How Pfizer Makes Its
     Millions of Covid-19 Vaccine Doses." *CNN*, April 2, 2021. https://www.
     cnn.com/2021/03/31/health/Pfizer-vaccine-manufacturing/index.html

42. Kansteiner, Fraiser, and Eric Sagonowsky. "What Does It Take
     to Supply COVID-19 Vaccines Across the Globe? Here's How
     the Leading Players Are Working It." Fierce Pharma, March
     3, 2021. https://www.fiercepharma.com/special-reports/
     vaccine-supply-chains-holding-line-against-Covid-19

43. Bill Bostock, "Even if a successful coronavirus vaccine is developed,
     billions could struggle to access it because of a global shortage of

glass vials," Insider, May 8, 2020. https://www.businessinsider.com/coronavirus-vaccine-glass-vial-shortage-could-delay-global-rollout-2020-5

44. Owermohle, "The 'Biggest Challenge' Won't Come Until After a Coronavirus Vaccine Is Found."

45. Sheffi, "The New (Ab)Normal: Reshaping Business and Supply Chain Strategy Beyond Covid-19," pp. 70–72.

46. Burger, Ludwig, and Matthias Blamont. "Bottlenecks? Glass Vial Makers Prepare for COVID-19 Vaccine." Reuters, June 12, 2020. https://www.reuters.com/article/us-health-coronavirus-schott-exclusive/exclusive-bottlenecks-glass-vial-makers-prepare-for-Covid-19-vaccine-idUSKBN23J0SN

47. Rowland, "A Race Is On to Make Enough Small Glass Vials to Deliver Coronavirus Vaccine Around the World."

48. LaFraniere, Sharon, Noah Weiland, and Sheryl Gay Stolberg. "F.D.A. Agrees Moderna Can Increase Vaccine Supply in Each Vial." New York Times, February 12, 2021. https://www.nytimes.com/2021/02/12/us/politics/Moderna-coronavirus-vaccine-supply.html

49. Thomas, Katie. "Hospitals Discover a Surprise in Their Vaccine Deliveries: Extra Doses." New York Times, December 16, 2020. https://www.nytimes.com/2020/12/16/health/Covid-Pfizer-vaccine-extra-doses.html

50. The White House and Presidented Joseph R. Biden, Jr. "National Strategy for the COVID-19 Response and Pandemic Preparedness." January 21, 2021. https://www.whitehouse.gov/wp-content/uploads/2021/01/National-Strategy-for-the-COVID-19-Response-and-Pandemic-Preparedness.pdf

51. Klein, Betsy. "Biden Administration to Use Defense Production Act for Pfizer Supplies, At-Home Tests and Masks." CNN, February 5, 2021. https://www.cnn.com/2021/02/05/politics/defense-production-act-Pfizer-masks/index.html

52. Delbert, Caroline. "With Few Willing to Fly, Airliners Are Transforming Into Cargo Planes." Popular Mechanics, March 24, 2020, https://www.popularmechanics.com/flight/airliners/a31914424/passenger-airliners-cargo-planes/.

53. Author interview with Meri Stevens, Worldwide Vice President, Consumer Health Supply Chain and Deliver, Johnson & Johnson, June 4, 2020. (On the day of the interview, J&J announced that Stevens was

given the additional responsibility to lead the company's consumer health supply chain.)

54. Author interview with Meri Stevens, Worldwide Vice President Supply Chain – Consumer Health and Deliver at Johnson & Johnson, MIT CTL webinar/interview, March 17, 2021

55. Ibid.

56. BioNTech AG and Pfizer, Inc. "BioNTech Signs Collaboration Agreement with Pfizer to Develop mRNA-based Vaccines for Prevention of Influenza." [Press release] August 16, 2018. https://BioNTech.de/sites/default/files/2019-08/20180816_BioNTech-Signs-Collaboration-Agreement-with-Pfizer.pdf

57. Pfizer, Inc. "Pfizer Reports Fourth-Quarter and Full-Year 2019 Results." [Press release]. January 28, 2020. https://investors.Pfizer.com/investor-news/press-release-details/2020/Pfizer-REPORTS-FOURTH-QUARTER-AND-FULL-YEAR-2019-RESULTS/default.aspx

58. BioNTech AG. "BioNTech Announces Full Year 2019 Financial Results and Corporate Update." [Press release]. March 31, 2020. https://www.globenewswire.com/en/news-release/2020/03/31/2008996/0/en/-BioNTech-Announces-Full-Year-2019-Financial-Results-and-Corporate-Update.html

59. Pfizer, Inc. "Pfizer and BioNTech to Co-Develop Potential COVID-19 Vaccine." [Press release]. March 17, 2020. https://investors.Pfizer.com/investor-news/press-release-details/2020/Pfizer-and-BioNTech-to-Co-Develop-Potential-Covid-19-Vaccine/default.aspx

60. Paris, Costas. "Supply-Chain Obstacles Led to Last Month's Cut to Pfizer's Covid-19 Vaccine-Rollout Target." *Wall Street Journal,* December 3, 2020. https://www.wsj.com/articles/Pfizer-slashed-its-Covid-19-vaccine-rollout-target-after-facing-supply-chain-obstacles-11607027787

61. Sealy, "Manufacturing Moonshot: How Pfizer Makes Its Millions of Covid-19 Vaccine Doses."

62. Griffin, Riley. "J&J Looks for Partners to Ramp Up Supply of Covid-19 Vaccine." *Bloomberg,* March 1, 2021. https://www.bloomberg.com/news/articles/2021-03-01/j-j-looks-for-partners-to-ramp-up-supply-of-Covid-19-vaccine

63. Levine, Hallie. "From Lab to Vaccine Vial: The Historic Manufacturing Journey of Johnson & Johnson's Janssen COVID-19 Vaccine." Johnson & Johnson. March 3, 2021. https://www.jnj.com/innovation/making-johnson-johnson-janssen-Covid-19-vaccine

64. Author interview with Meri Stevens, Worldwide Vice President Supply Chain – Consumer Health and Deliver at Johnson & Johnson, MIT CTL webinar/interview, March 17, 2021

65. Peters, Adele. "Inside one of the new, quick-build factories making the Moderna vaccine." *Fast Company*, December 21, 2020. https://www. fastcompany.com/90588372/inside-one-of-the-new-quick-build-factories-making-the-Moderna-vaccine

66. Mullin, "Pfizer, Moderna Ready Vaccine Manufacturing Networks."

67. Peters, "Inside one of the new, quick-build factories making the Moderna vaccine."

68. Author interview with Marcello Damiani, Chief Digital and Operational Excellence Officer at Moderna, April 22, 2021

69. Ustinova, "In the Thick of the 'Herculean' Vaccine Push."

70. Author interview with Marcello Damiani, Chief Digital and Operational Excellence Officer at Moderna, April 22, 2021

71. Bourla, Albert. "Moving at the Speed of Science." Pfizer, Inc. Accessed September 29, 2021. https://www.Pfizer.com/news/hot-topics/moving_at_the_speed_of_science

72. Weise, Elizabeth. "US Cuts $1.95 Billion Deal With Pfizer for 100 Million Doses of COVID-19 Vaccine." *USA Today*, July 22, 2020. https://www.usatoday.com/story/news/health/2020/07/22/us-pays-1-95-billion-100-million-doses-Pfizer-Covid-19-vaccine/5489964002/

73. Paris, "Supply-Chain Obstacles Led to Last Month's Cut to Pfizer's Covid-19 Vaccine-Rollout Target."

74. Arthur, Rachel. "Pfizer COVID-19 vaccine: 2 billion doses and $15bn sales expected in 2021." BioPharma Reporter, February 2, 2021. https://www.biopharma-reporter.com/Article/2021/02/02/Pfizer-BioNTech-to-produce-2-billion-Covid-19-vaccine-doses-in-2021

75. Higgins-Dunn, Noah. "BioNTech, With Partner Pfizer, On Track to Make 3B COVID Vaccine Doses in 2021, CEO Says." Fierce Pharma, May 4, 2021. https://www.fiercepharma.com/pharma/thanks-to-revved-up-manufacturing-BioNTech-ceo-estimates-3b-covid-vaccine-doses-2021

76. Parsons, Lucy. "BioNTech on Course to Produce Three Billion Doses of Pfizer-Partnered COVID-19 Vaccine in 2021." PM Live, May 5, 2021. https://www.pmlive.com/pharma_news/BioNTech_on_course_to_produce_three_billion_doses_of_Pfizer-partnered_covid-19_vaccine_in_2021_1369187

77. Pagliarulo, Ned. "Vaccine Factories Churn Out Millions More Doses, Speeding US Rollout of Coronavirus Shots." https://www. biopharmadive.com/news/vaccine-factories-churn-out-millions-more-doses-speeding-us-rollout-of-cor/596561/

78. Kresge, Naomi, and Riley Griffin. "Pfizer to Cut Vaccine Shipments as Belgian Factory Renovated." Bloomberg, January 15, 2021. https://www. bloomberg.com/news/articles/2021-01-15/Pfizer-to-cut-covid-vaccine-deliveries-as-it-renovates-factory

79. "Coronavirus vaccine delays halt Pfizer jabs in parts of Europe." *BBC News*, January 20, 2021. https://www.bbc.com/news/world-europe-55765556

80. GlobalData Healthcare. "Manufacturing disruptions delay Covid-19 vaccine distribution." Pharmaceutical Technology, January 28, 2021. https://www.pharmaceutical-technology.com/comment/Covid-19-vaccine-manufacturing-disruptions/

81. Arthur, Rachel. "Pfizer COVID-19 vaccine: 2 billion doses and $15bn sales expected in 2021."

82. Kresge, Naomi. "BioNTech Raises Covid Vaccine Target to 2.5 Billion Doses." Bloomberg, March 30, 2021. https://www.bloomberg.com/news/articles/2021-03-30/BioNTech-raises-2021-covid-vaccine-target-to-2-5-billion-doses

83. Higgins-Dunn, "BioNTech, With Partner Pfizer, On Track to Make 3B COVID Vaccine Doses in 2021, CEO Says."

84. Paris, Costas, and Jared S. Hopkins. "Pfizer Sets Up Its 'Biggest Ever' Vaccination Distribution Campaign." *Wall Street Journal*, October 21, 2020. https://www.wsj.com/articles/Pfizer-sets-up-its-biggest-ever-vaccination-distribution-campaign-11603272614

85. Author interview with Phillip Sharp, Nobel Prize-winning biochemist and Institute Professor at MIT, March 17, 2021

86. Mullin, "Pfizer, Moderna Ready Vaccine Manufacturing Networks."

87. Pfizer, Inc. "Pfizer-BioNTech COVID-19 Vaccine U.S. Distribution Fact Sheet." [Press release]. Author, November 2020. https://www.Pfizer.com/news/hot-topics/covid_19_vaccine_u_s_distribution_fact_sheet)

88. Packaging Europe. "Softbox Supports Pfizer in Global Cold Chain Distribution of COVID-19 Vaccine." Author, March 11, 2021. https://packagingeurope.com/softbox-supports-Pfizer-in-global-cold-chain-distribution-of-COVID-vaccine/

89.   Lamers, Vanessa. "FAQs on the Pfizer Vaccine: Shipping, Handling, Preparation, and Administration." PHF Pulse [blog]. December 15, 2020. http://www.phf.org/phfpulse/Pages/FAQs_on_the_Pfizer_Vaccine_Shipping_Handling_Preparation_and_Administration.aspx

90.   Bailey, Joanna. "Pfizer Increases Dry Ice Life Span Improving Vaccine Capacity On Boeing Aircraft." Simple Flying, December 15, 2020. https://simpleflying.com/boeing-aircraft-vaccine-capacity-increased/

91.   Bourla, Albert. "Distributing Our COVID-19 Vaccine to the World." Pfizer, Inc. Accessed September 29, 2021. https://www.Pfizer.com/news/hot-topics/distributing_our_covid_19_vaccine_to_the_world

92.   Butcher, Mike. "Iceland's Controlant, with $50M backing, emerges as key player in Cold Chain for COVID-19 vaccine." TechCrunch, December 15, 2020. https://techcrunch.com/2020/12/15/icelands-controlant-with-50m-backing-emerges-as-key-player-in-cold-chain-for-Covid-19-vaccine/

93.   Controlant. "Controlant now providing monitoring and Supply Chain Visibility for Pfizer-BioNTech COVID-19 Vaccine distribution and storage." [Press release]. Author: December 15, 2020. https://controlant.com/blog/2020/controlant-now-providing-monitoring-and-supply-chain-visibility-for-Pfizer-BioNTech-Covid-19-vaccine-distribution-and-storage/

94.   Goldhill, Olivia. "Pfizer Decision to Turn Off Temperature Sensors Forced Scramble to Ensure Covid-19 Vaccines Kept Ultra-Cold." Stat, December 17, 2020. https://www.statnews.com/2020/12/17/Pfizer-decision-to-turn-off-temperature-sensors-forced-scramble-to-ensure-Covid-19-vaccines-kept-cold/

95.   Meredith, Sam. "Pfizer, BioNTech Say Covid Vaccine Is More Than 90% Effective — 'Great Day for Science and Humanity'." CNBC, November 9, 2020. https://www.cnbc.com/2020/11/09/covid-vaccine-Pfizer-drug-is-more-than-90percent-effective-in-preventing-infection.html

96.   Sigalos, MacKenzie. "Pfizer's Covid Vaccine Is Now Shipping. Here's How the U.S. Plans to Deliver It." CNBC, December 12, 2020. https://www.cnbc.com/2020/12/12/how-fedex-ups-plan-to-distribute-fda-approved-covid-vaccine-when-will-you-get-the-coronavirus-vaccine.html

97.   Katz, Benjamin, Doug Cameron and Alison Sider. "How Airlines Are Rushing to Deliver Covid-19 Vaccines." *Wall Street Journal*, December 16, 2020. https://www.wsj.com/articles/Covid-19-vaccines-are-coming-airlines-are-rushing-to-deliver-them-11608147229

98. Brett, Damian. "US cargo carriers fly US Covid-19 vaccine shipments following FDA approval." Air Cargo News, December 14, 2020. https://www.aircargonews.net/sectors/pharma-logistics/fedex-united-and-ups-fly-us-Covid-19-vaccine-shipments-following-fda-approval/

99. Kulisch, Eric. "FedEX, UPS trucks depart with first Pfizer COVID vaccines." American Shipper, December 13, 2020. https://www.freightwaves.com/news/fedex-ups-trucks-depart-with-first-Pfizer-covid-vaccines

References for Chapter 3

1.   Bloch, Andrew. Twitter post, January 5, 2021, 7:26 AM. https://twitter.
     com/andrewbloch/status/1346432760920084480?lang=en

2.   Holland, Frank. "Amazon is delivering nearly two-thirds of its own
     packages as e-commerce continues pandemic boom." CNBC, August 13,
     2020. https://www.cnbc.com/2020/08/13/amazon-is-delivering-nearly-
     two-thirds-of-its-own-packages.html

3.   Centers for Disease Control and Prevention. "People with Certain
     Medical Conditions." Last updated August 20, 2021. https://www.cdc.
     gov/coronavirus/2019-ncov/need-extra-precautions/people-with-
     medical-conditions.html

4.   Weise, Elizabeth. "US Cuts $1.95 Billion Deal With Pfizer for 100 Million
     Doses of COVID-19 Vaccine." *USA Today,* July 22, 2020. https://www.
     usatoday.com/story/news/health/2020/11/18/Pfizer-Covid-19-vaccine-
     dry-ice-sales-storage/6281859002/

5.   Our World in Data. "Share of the Population Fully Vaccinated Against
     COVID-19." Accessed September 30, 2021. https://ourworldindata.org/
     grapher/share-people-fully-vaccinated-covid?country=USA~ISR~Europe

6.   Times of Israel staff and agencies. "Israel said to ink deal with Moderna
     for potential purchase of COVID-19 vaccine." The Times of Israel, June
     16, 2020. https://www.timesofisrael.com/israel-said-to-ink-deal-with-
     Moderna-for-potential-purchase-of-covid-19-vaccine/

7.   Bassist, Rina. "Netanyahu says Israel signed agreement with Pfizer for
     coronavirus vaccine." Al-Monitor, November 13, 2020. https://www.
     al-monitor.com/originals/2020/11/israel-benjamin-netanyahu-Pfizer-
     Moderna-coronavirus-vaccine.html

8.   Jaffe-Hoffman, Maayan. "Edelstein: Israel to Sign Deal for Pfizer Vaccine
     on Friday." November 12, 2020. https://www.jpost.com/breaking-news/
     pmo-netanyahu-spoke-with-Pfizer-ceo-again-working-on-vaccine-
     contract-648852

9.   Winer, Stuart. "Israel said to be paying average of $47 per person for
     Pfizer, Moderna vaccines." *Times of Israel,* January 12, 2021. https://www.
     timesofisrael.com/israel-said-to-be-paying-average-of-47-per-person-
     for-Pfizer-Moderna-vaccines/

10. CountryEconomy.com. "Israel GDP – Gross Domestic Product. Accessed
     September 30, 2021. https://countryeconomy.com/gdp/israel

11.	Bureau of Economic Analysis. "Gross Domestic Product by State, 1st Quarter 2021." [Press release]. U.S. Department of Commerce. June 25, 2021. https://www.bea.gov/news/2021/gross-domestic-product-state-1st-quarter-2021

12.	Eurostat. "Euroindicators." June 8, 2021. https://ec.europa.eu/eurostat/documents/2995521/11563119/2-08062021-AP-EN.pdf/eead4cc5-f4f2-a087-9ded-a1b15bf2394a?t=1623140343558

13.	Government of Israel. "Real-World Epidemiological Evidence Collaboration Agreement." Accessed September 30, 2021. https://govextra.gov.il/media/30806/11221-moh-Pfizer-collaboration-agreement-redacted.pdf

14.	Rasgon, Adam. "Israel Reaches a Deal With Pfizer for Enough Vaccine to Inoculate All Its Population Over 16 by the End of March." *New York Times*, January 7, 2021. https://www.nytimes.com/2021/01/07/world/israel-reaches-a-deal-with-Pfizer-for-enough-vaccine-to-inoculate-all-its-population-over-16-by-the-end-of-march.html

15.	Global Labor Organization. "Israel's Vaccination Success Story. Interview with Professor Gil S. Epstein, Bar Ilan University." GLO, February 26, 2021. https://glabor.org/israels-vaccination-success-story-interview-with-professor-gil-s-epstein-bar-ilan-university/

16.	Author interview with Gil Epstein, Professor of Economics and Dean of Social Science at Bar-Ilan University, April 18, 2021.

17.	"Israel Receives Initial Shipment of Pfizer Coronavirus Vaccine." CNBC, December 9, 2020. https://www.cnbc.com/2020/12/09/israel-to-receive-initial-shipment-of-Pfizer-coronavirus-vaccine.html

18.	Author interview with Gil Epstein, Professor of Economics and Dean of Social Science at Bar-Ilan University, April 18, 2021.

19.	World Health Organization. "WHO Sage Roadmap for Prioritizing Uses of COVID-19 Vaccines in the Context of Limited Supply." WHO, November 13, 2020. https://cdn.who.int/media/docs/default-source/immunization/sage/covid/sage-prioritization-roadmap-Covid-19-vaccines_31a59ccd-1fbf-4a36-a12f-73344134e49d.pdf?sfvrsn=bf227443_36&download=true

20.	Gross, Judah Ari. "IDF declares: We are first military in the world with herd immunity." *Times of Israel*, March 11, 2021. https://www.timesofisrael.com/idf-declares-we-are-first-military-in-the-world-with-herd-immunity/

21. Author interview with Gil Epstein, Professor of Economics and Dean of Social Science at Bar-Ilan University, April 18, 2021.

22. Miller, Lee J., and Wei Lu. "Asia Trounces U.S. in Health-Efficiency Index Amid Pandemic." Bloomberg, December 17, 2020. https://www.bloomberg.com/news/articles/2020-12-18/asia-trounces-u-s-in-health-efficiency-index-amid-pandemic

23. Author conversation with Esti Marian, May 23, 2021.

24. Centers for Disease Control & Prevention. "Pfizer-BioNTech COVID-19 Vaccine Storage and Handling Summary." August 24, 2021. https://www.cdc.gov/vaccines/covid-19/info-by-product/Pfizer/downloads/storage-summary.pdf

25. Nevett, Joshua. "Covid-19 Vaccines: Is It OK to Get a Leftover Jab?" *BBC News*, January 28, 2021. https://www.bbc.com/news/world-55841017

26. Our World in Data. "Daily COVID-19 Vaccine Doses Administered." Accessed September 30, 2021. https://ourworldindata.org/explorers/coronavirus-data-explorer?zoomToSelection=true&time=2020-03-01..latest&pickerSort=desc&pickerMetric=total_vaccinations_per_hundred&Metric=Vaccine+doses&Interval=7-day+rolling+average&Relative+to+Population=true&Align+outbreaks=false&country=USA~GBR~ISR~European+Union~OWID_WRL~CHL

27. WorldOMeter. "Coronavirus: Israel." Accessed September 30, 2021. https://www.worldometers.info/coronavirus/country/israel/#coronavirus-deaths-log

28. Launch & Scale Speedometer. "Tracking COVID-19 Vaccine Purchases Across the Globe." Accessed September 30, 2021. https://launchandscalefaster.org/covid-19/vaccinepurchases

29. Ellyatt, Holly. "India is the home of the world's biggest producer of Covid vaccines. But it's facing a major internal shortage." CNBC, May 5, 2021. https://www.cnbc.com/2021/05/05/why-covid-vaccine-producer-india-faces-major-shortage-of-doses.html

30. Dunn, Lauren. "Delayed doses, unfulfilled requests and last-minute allocations: Inside the vaccine rollout." NBC News, January 29, 2021. https://www.nbcnews.com/health/health-news/delayed-doses-unfulfilled-requests-last-minute-allocations-inside-vaccine-rollout-n1256072

31. Krouse, Sarah, Brianna Abbott, and Jared S. Hopkins. "Behind America's Botched Vaccination Rollout: Fragmented Communication, Misallocated Supply." *Wall Street Journal*, February 18, 2021. https://www.wsj.com/

articles/behind-americas-botched-vaccination-rollout-fragmented-communication-misallocated-supply-11613663012

32.   Johnson, Carla K., and Nicky Forster. "A rapid COVID-19 vaccine rollout backfired in some US states." AP News, March 21, 2021. https://apnews.com/article/rapid-Covid-19-vaccine-rollout-backfire-some-states-459feb d1a33bf8909bfaa88ce3f43159

33.   Bergal, Jenni. "Vaccine Signups Cater to the Tech-Savvy, Leaving Out Many." The Pew Charitable Trusts, February 10, 2021. https://www.pewtrusts.org/en/research-and-analysis/blogs/stateline/2021/02/10/vaccine-signups-cater-to-the-tech-savvy-leaving-out-many

34.   Johnson and Forster, "A rapid COVID-19 vaccine rollout backfired in some US states."

35.   Dunn, "Delayed doses, unfulfilled requests and last-minute allocations: Inside the vaccine rollout."

36.   Gillies, Rob. "Canada vaccine panel recommends 4 months between COVID doses." Associated Press/ABC News, March 3, 2021. https://abcnews.go.com/International/wireStory/canada-vaccine-panel-recommends-months-covid-doses-76239947

37.   Pancevski, Bojan. "U.K. Delays Second Covid-19 Vaccine Dose as Europe Ponders How to Speed Up Immunization." *Wall Street Journal*, December 30, 2020. https://www.wsj.com/articles/u-k-delays-second-Covid-19-vaccine-dose-as-europe-ponders-how-to-speed-up-immunization-11609334172

38.   Wu, Katherine J., and Rebecca Robbins. "In Europe, more countries delay second vaccine doses or mull plans to do so." *New York Times*, January 4, 2021. Updated January 5, 2021. https://www.nytimes.com/2021/01/04/world/second-covid-vaccine-delay.html

39.   Robertson, John F.R., Herb F. Sewell, and Marcia Stewart. "Delayed Second Dose of the BNT162b2 Vaccine: Innovation or Misguided Conjecture?" *Lancet*, February 19, 2021. https://www.thelancet.com/journals/lancet/article/PIIS0140-6736(21)00455-4/fulltext

40.   Public Health England. "Vaccines Highly Effective Against B.1.617.2 Variant After 2 Doses." [Press release]. May 22, 2021. https://www.gov.uk/government/news/vaccines-highly-effective-against-b-1-617-2-variant-after-2-doses

41.   Schuster-Bruce, Catherine. "How Much Protection You Get From One Shot of the Pfizer, AstraZeneca, and Moderna Vaccines, According to the Best Available Data." Insider, July 9, 2021. https://www.businessinsider.

com/covid-vaccine-one-shot-effectiveness-Pfizer-Moderna-AstraZeneca-vaccines-dose-2021-3

42. Lewis, Dyani. "Mix-and-Match COVID Vaccines: The Case Is Growing, But Questions Remain." *Nature*, 595 (2021): 344–345. https://www.nature.com/articles/d41586-021-01805-2

43. Winfield Cunningham, Paige, with Alexandra Ellerbeck. "The Health 202: How West Virginia beat other states in administering coronavirus vaccines." *Washington Post*, February 4, 2021. https://www.washingtonpost.com/politics/2021/02/04/health-202-how-west-virginia-beat-other-states-administering-coronavirus-vaccines/

44. Ellyatt, Holly. "Israel's Covid Vaccine Rollout Is the Fastest in the World — Here Are Some Lessons for the Rest of Us." CNBC, January 8, 2021. https://www.cnbc.com/2021/01/07/israels-covid-vaccine-rollout-is-the-fastest-in-the-world.html

45. Rosen, Bruce, Ruth Waitzberg, and Avi Israeli. "Israel's Rapid Rollout of Vaccinations for COVID-19." *Israel Journal of Health Policy Research* 10 (January 26, 2021): 6. https://ijhpr.biomedcentral.com/articles/10.1186/s13584-021-00440-6

46. Kirby, Jen. "Why Israel Is Leading the World in Vaccinating Its Population." Vox, January 14, 2021. https://www.vox.com/2021/1/14/22215896/israel-vaccine-coronavirus-Pfizer-netanyahu

47. "Everyone Has a Plan Until They Get Punched in the Mouth." [Blog]. Commit.Works. Accessed September 30, 2021. https://www.commit.works/everyone-has-a-plan-until-they-get-punched-in-the-mouth/

48. Eisenhower, Dwight D. [attributed]. "Plans Are Worthless, But Planning Is Everything." Quote Investigator. Accessed September 30, 2021. https://quoteinvestigator.com/2017/11/18/planning/

49. Williams, Michelle. "Massachusetts COVID vaccine website crashes the morning residents 65+, individuals with multiple health conditions become eligible for appointments." MassLive, February 18, 2021. https://www.masslive.com/coronavirus/2021/02/massachusetts-covid-vaccine-website-crashes-the-morning-residents-65-individuals-with-multiple-health-conditions-become-eligible-for-appointments.html

50. Chen, Angus. "What Went Wrong With The 'Tiny Nonprofit' Behind The State's Beleaguered Vaccine Site." WBUR, February 25, 2021. https://www.wbur.org/commonhealth/2021/02/25/what-went-wrong-with-the-tiny-nonprofit-behind-the-states-beleaguered-vaccine-site

51. Gardizy, Anissa, Kay Lazar, and Adam Vaccaro. "Massachusetts Spent 20 Years Refining Its Own Mass Vaccination Plan. Then It Looked Elsewhere." *Boston Globe*, March 4, 2021. https://www.bostonglobe.com/2021/03/05/nation/massachusetts-spent-20-years-fine-tuning-its-own-mass-vaccination-plan-then-it-looked-elsewhere

52. Massachusetts Department of Public Health. "Massachusetts' COVID-19 Vaccination Phases." Mass.gov. Accessed September 30, 2021. https://www.mass.gov/info-details/massachusetts-Covid-19-vaccination-phases

53. Barry, Ellen. "Will Massachusetts's Vaccine Buddy System Work? Well, It's Worth a Shot." *New York Times*, February 12, 2021. Updated February 22, 2021. https://www.nytimes.com/2021/02/12/us/covid-vaccine-caregivers-massachusetts.html

54. Goralnick, Eric, Christoph Kaufmann, and Atul A. Gawande. "Mass-Vaccination Sites — An Essential Innovation to Curb the Covid-19 Pandemic." *New England Journal of Medicine*, 384 (2021): 367. https://www.nejm.org/doi/full/10.1056/NEJMp2102535

55. Herwick, Edgar B. III. "Massachusetts' New Pre-Registration System For COVID Vaccination: An Explainer." WGBH News, March 12, 2021. https://www.wgbh.org/news/local-news/2021/03/12/massachusetts-new-pre-registration-system-for-covid-vaccination-an-explainer

56. McPhillips, Deidre, Madeline Holcombe, and Jason Hanna. "3 States Have Already Reached Biden's New Vaccination Goal, But Vaccine Hesitancy May Make it Challenging for Others." CNN, May 11, 2021. https://www.cnn.com/2021/05/05/health/us-coronavirus-wednesday/index.html

57. Sun, Lena H. "Who should get a coronavirus vaccine first?" *Washington Post*, July 29, 2020. https://www.washingtonpost.com/health/2020/07/29/covid-vaccine-essential-workers-high-risk-populations/

58. Ndugga, Nambi, Latoya Hill, and Samantha Artiga. "Latest Data on COVID-19 Vaccinations by Race/Ethnicity." Kaiser Family Foundation, September 22, 2021. https://www.kff.org/coronavirus-covid-19/issue-brief/latest-data-on-covid-19-vaccinations-race-ethnicity

59. Huntsman, Anna. "Ohio's Amish Suffered a Lot From Covid, but Vaccines Are Still a Hard Sell." KHN, April 28, 2021. https://khn.org/news/article/ohios-amish-suffered-a-lot-from-covid-but-vaccines-are-still-a-hard-sell/

60. Rothstein, Mark A., and Christine N. Coughlin. "Undocumented Immigrants and Covid-19 Vaccination." The Hastings

Center, March 8, 2021. https://www.thehastingscenter.org/undocumented-immigrants-and-Covid-19-vaccination/

61. Beckhusen, Julia. "Women More Likely to Have Multiple Jobs." U.S. Census Bureau. June 18, 2019. https://www.census.gov/library/stories/2019/06/about-thirteen-million-united-states-workers-have-more-than-one-job.html

62. Ndugga, Nambi, Samantha Artiga, and Olivia Pham. "How are States Addressing Racial Equity in COVID-19 Vaccine Efforts?" Kaiser Family Foundation, March 10, 2021. https://www.kff.org/racial-equity-and-health-policy/issue-brief/how-are-states-addressing-racial-equity-in-Covid-19-vaccine-efforts

63. Hopkins, Jared S. "To Vaccinate Against Covid-19, U.S. Enlists Pharmacy Chains." *Wall Street Journal*, October 30, 2020. https://www.wsj.com/articles/to-vaccinate-against-covid-19-u-s-enlists-pharmacy-chains-11604107495

64. Japsen, Bruce. "In Boost To CVS And Walgreens, U.S. Expands Covid-19 Vaccination Powers To Pharmacy Techs." *Forbes*, November 2, 2020. https://www.forbes.com/sites/brucejapsen/2020/11/02/in-boost-to-cvs-and-walgreens-us-expands-Covid-19-vaccination-powers-to-pharmacy-techs/?sh=3a041734239b

65. Centers for Disease Control and Prevention. "Understanding the Federal Retail Pharmacy Program for COVID-19 Vaccination." Last reviewed: September 23, 2021. https://www.cdc.gov/vaccines/Covid-19/retail-pharmacy-program/index.html

66. Burns Loeb, Tamra, AJ Adkins-Jackson, and Arleen F. Brown. "No Internet, No Vaccine: How Lack of Internet Access Has Limited Vaccine Availability for Racial and Ethnic Minorities." The Conversation, February 8, 2021. https://theconversation.com/no-internet-no-vaccine-how-lack-of-internet-access-has-limited-vaccine-availability-for-racial-and-ethnic-minorities-154063

67. Bergal, "Vaccine Signups Cater to the Tech-Savvy, Leaving Out Many."

68. O'Brien, Matt, and Candice Choi. "Explainer: Meet the Vaccine Appointment Bots, and Their Foes." AP News, February 25, 2021. https://apnews.com/article/public-health-new-jersey-media-social-media-coronavirus-pandemic-5590b7f0cdd5d649f5f52d8c26e48112

69. Our World in Data. "COVID-19 Vaccine Doses Administered Per 100 People." Accessed September 30, 2021. https://ourworldindata.org/

grapher/covid-vaccination-doses-per-capita?country=CHN~ISR~USA~E
uropean+Union~Africa~South+America~IND

70. Launch & Scale Speedometer. "Tracking COVID-19 Vaccine Purchases
    Across the Globe."

71. Grainger, Matt, and Sara Dransfield. "Rich Nations Vaccinating
    One Person Every Second While Majority of the Poorest Nations
    Are Yet to Give a Single Dose." UNAIDS, March 10, 2021. https://
    www.unaids.org/en/resources/presscentre/featurestories/2021/
    march/20210310_Covid-19-vaccines

72. Roy, Rajesh, and Vibhuti Agarwal. "India Suspends Covid-19 Vaccine
    Exports to Focus on Domestic Immunization." *Wall Street Journal,*
    March 25, 2021. https://www.wsj.com/articles/india-suspends-
    Covid-19-vaccine-exports-to-focus-on-domestic-immunization-
    11616690859?mod=article_inline

73. Norman, Laurence, and Jenny Strasburg. "Vaccine Fight Between EU and
    U.K. Threatens to Escalate." *Wall Street Journal,* March 25, 2021. https://
    www.wsj.com/articles/vaccine-fight-between-eu-and-u-k-threatens-to-
    escalate-11616444756?mod=article_inline

74. "The Many Guises of Vaccine Nationalism." *Economist,* March 13, 2021.
    https://www.economist.com/finance-and-economics/2021/03/11/
    the-many-guises-of-vaccine-nationalism

75. Federal Emergency Management Agency. "Defense Production Act."
    Accessed September 30, 2021. https://www.fema.gov/disasters/
    defense-production-act

76. Federal Emergency Management Agency. "Applying the Defense
    Production Act." [Press release]. January 26, 2021. https://www.fema.
    gov/press-release/20210420/applying-defense-production-act

77. Martell, Allison, and Euan Rocha. "How the U.S. Locked
    Up Vaccine Materials Other Nations Urgently Need."
    Reuters, May 7, 2021. https://www.reuters.com/article/
    us-health-coronavius-vaccines-dpa-insigh-idCAKBN2CO1I8

78. Weintraub, Arlene. "Feds Rebuff Pfizer's Pleas to Speed Up Supplies of
    COVID-19 Vaccine Raw Materials: Reports." Fierce Pharma, December
    15, 2020. https://www.fiercepharma.com/pharma/feds-rebuff-Pfizer-s-
    requests-for-speedier-supplies-covid-vaccine-raw-materials-reports

79. LaFraniere, Sharon, and Katie Thomas. "Pfizer Nears Deal With Trump
    Administration to Provide More Vaccine Doses." *New York Times,*

December 22, 2020. https://www.nytimes.com/2020/12/22/us/politics/-Pfizer-vaccine-doses.html

80.   Lupkin, Sydney. "Defense Production Act Speeds Up Vaccine Production." *NPR*, March 13, 2021. https://www.npr.org/sections/health-shots/2021/03/13/976531488/defense-production-act-speeds-up-vaccine-production

81.   Evenett, Simon J., Bernard Hoekman, Nadia Rocha, and Michele Ruta. "The Covid-19 Vaccine Production Club Will Value Chains Temper Nationalism?" World Bank Group Policy Research Working Paper 9565. http://documents1.worldbank.org/curated/en/244291614991534306/pdf/The-Covid-19-Vaccine-Production-Club-Will-Value-Chains-Temper-Nationalism.pdf

82.   Michalopoulos, Sarantis. "Pharma groups 'losing time' with EU export control mechanism." Euractiv, February 5, 2021. https://www.euractiv.com/section/coronavirus/news/pharma-groups-losing-time-with-eu-export-control-mechanism

83.   Michalopoulos, Sarantis. "EU's COVID-19 vaccine export ban risks global retaliation, warns pharma industry." Euractiv, January 29, 2021. https://www.euractiv.com/section/coronavirus/news/eus-Covid-19-vaccine-export-ban-risks-global-retaliation-warns-pharma-industry/

84.   Covid-19 Vaccine Tracker [website]. Accessed September 30, 2021. https://Covid-19.trackvaccines.org/vaccines/

85.   "Vaccine diplomacy boosts Russia's and China's global standing." *Economist,* April 29, 2021. https://www.economist.com/graphic-detail/2021/04/29/vaccine-diplomacy-boosts-russias-and-chinas-global-standing

86.   Smith, Alexander. "Russia and China Are Beating the U.S. at Vaccine Diplomacy, Experts Say." *NBC News,* April 2, 2021. https://www.nbcnews.com/news/world/russia-china-are-beating-u-s-vaccine-diplomacy-experts-say-n1262742

87.   Connors, Emma. "Vaccine Diplomacy a Threat to South China Sea Pushback." *Financial Review,* September 2, 2020. https://www.afr.com/world/asia/vaccine-diplomacy-a-threat-to-south-china-sea-pushback-20200901-p55raa

88.   Baraniuk, Chris. "Covid-19: What Do We Know About Sputnik V and Other Russian Vaccines?" *BMJ*, 372 (2021): n743. https://www.bmj.com/content/372/bmj.n743

89. Covid-19 Vaccine Tracker. "Gamaleya: Sputnik V." Accessed August 9, 2021. https://Covid-19.trackvaccines.org/vaccines/12/

90. Baraniuk, "Covid-19: What Do We Know About Sputnik V and Other Russian Vaccines?"

91. "Vaccine diplomacy boosts Russia's and China's global standing." *Economist*, April 29, 2021.

92. Ramos, Daniel, Aislinn Laing, and Cassandra Garrison. "Amid Scramble for COVID-19 Vaccine, Latin America Turns to Russia." Reuters, March 1, 2021. https://www.reuters.com/article/us-health-coronavirus-latam-russia-insig/amid-scramble-for-covid-19-vaccine-latin-america-turns-to-russia-idUSKCN2AT23J

93. Thakur, Dinesh. "India Is Suffering Immensely Under the Weight of Covid. Now Its Failures Are Threatening Much of the World." *Stat*, May 5, 2021. https://www.statnews.com/2021/05/05/india-vaccine-heist-shoddy-regulatory-oversight-imperil-global-vaccine-access/

94. Frayer, Lauren. "The World's Largest Vaccine Maker Took A Multimillion Dollar Pandemic Gamble." *NPR*, March 18, 2021. https://www.npr.org/sections/goatsandsoda/2021/03/18/978065736/indias-role-in-Covid-19-vaccine-production-is-getting-even-bigger

95. Kelemen, Michele. "Quad Leaders Announce Effort To Get 1 Billion COVID-19 Vaccines To Asia." *NPR*, March 12, 2021. https://www.npr.org/2021/03/12/976589457/quad-leaders-announce-effort-to-get-1-billion-Covid-19-vaccines-to-asia

96. Dhume, Sadanand. "India Beats China at Vaccine Diplomacy." *Wall Street Journal*, March 18, 2021. https://www.wsj.com/articles/india-beats-china-at-vaccine-diplomacy-11616086729

97. Dhar, Biswajit. "India's Vaccine Diplomacy." *Global Policy*, April 8, 2021. https://www.globalpolicyjournal.com/blog/08/04/2021/indias-vaccine-diplomacy

98. Pant, Harsh V., and Premesha Saha. "India's Vaccine Diplomacy Reaches Taiwan." *Foreign Policy*, April 20, 2021. https://foreignpolicy.com/2021/04/20/india-vaccine-diplomacy-china-taiwan/

99. Thakur, "India Is Suffering Immensely Under the Weight of Covid. Now Its Failures Are Threatening Much of the World."

100. Moutinho, Sofia, and Meredith Wadman. "Is Russia's COVID-19 Vaccine Safe? Brazil's Veto of Sputnik V Sparks Lawsuit Threat and Confusion." *Science*, April 30, 2021. https://www.sciencemag.org/news/2021/04/

russias-Covid-19-vaccine-safe-brazils-veto-sputnik-v-sparks-lawsuit-threat-and

101. Paraguassu, Lisandra, and Andrew Osborn. "Russian Vaccine Developer to Sue Brazilian Regulator for Defamation." Reuters, April 29, 2021. https://www.reuters.com/business/healthcare-pharmaceuticals/russian-vaccine-developer-sue-brazilian-regulator-defamation-2021-04-29/

102. Cook, Lorne. "EU Report Takes Aim at Russia Over Vaccine Fake News." ABC News/Associated Press, April 28, 2021. https://abcnews.go.com/Health/wireStory/eu-report-takes-aim-russia-vaccine-fake-news-77371470

103. Baraniuk, "Covid-19: What Do We Know About Sputnik V and Other Russian Vaccines?"

104. Cohen, Jon. "Russia's Claim of a Successful COVID-19 Vaccine Doesn't Pass the 'Smell Test,' Critics Say." Science, November 11, 2020. https://www.sciencemag.org/news/2020/11/russia-s-claim-successful-Covid-19-vaccine-doesn-t-pass-smell-test-critics-say

105. Moutinho and Wadman, "Is Russia's COVID-19 Vaccine Safe? Brazil's Veto of Sputnik V Sparks Lawsuit Threat and Confusion."

106. Higgins, Andrew. "Russian Attempts to Expand Sputnik Vaccine Set Off Discord in Europe." New York Times, May 2, 2021. https://www.nytimes.com/2021/05/02/world/europe/russia-slovakia-europe-coronavirus-sputnik-vaccine.html

107. "Sinovac: Brazil Results Show Chinese Vaccine 50.4% Effective." BBC News, January 13, 2021. https://www.bbc.com/news/world-latin-america-55642648

108. Launch & Scale Speedometer. "Tracking COVID-19 Vaccine Purchases Across the Globe: Tab. 1.3, Overview of Confirmed & Potential Deals." Accessed September 30, 2021. https://launchandscalefaster.org/covid-19/vaccinepurchases

109. Wee, Sui-Lee, and Ernesto Londoño. "Disappointing Chinese Vaccine Results Pose Setback for Developing World." New York Times, January 13, 2021. https://nyti.ms/3oFAtMT

110. "UAE, Bahrain Make Pfizer/BioNTech Shot Available to Those Who Got Sinopharm Vaccine." Reuters, June 3, 2021. https://www.aol.com/news/uae-bahrain-Pfizer-BioNTech-shot-163313383.html

111. Nair, Adveith. "UAE, Bahrain Plan Sinopharm Booster Shots Amid Efficacy Concerns." Bloomberg, May 18, 2021. Updated on May

19, 2021. https://www.bloomberg.com/news/articles/2021-05-18/
uae-to-offer-third-sinopharm-booster-shot-amid-efficacy-concerns

112.   LaFraniere, Sharon, and Noah Weiland. "For Biden, a New Virus
       Dilemma: How to Handle a Looming Glut of Vaccine." *New York Times*,
       March 26, 2021. https://www.nytimes.com/2021/03/26/us/biden-
       coronavirus-vaccine.html

113.   Brueninger, Kevin. "Biden to say U.S. will send 20 million Pfizer,
       Moderna or J&J vaccine doses abroad by end of June." CNBC, May 17,
       2021. https://www.cnbc.com/2021/05/17/biden-to-say-us-will-send-20-
       million-Pfizer-Moderna-or-jj-vaccine-doses-abroad-by-end-of-june.
       html

114.   Pettypiece, Shannon, and Lauren Egan. "Biden administration to share
       millions of AstraZeneca vaccine doses with Canada, Mexico." *NBC News*,
       March 18, 2021. https://www.nbcnews.com/politics/white-house/biden-
       administration-share-millions-AstraZeneca-vaccine-doses-canada-
       mexico-n1261425

115.   LaFraniere, Sharon, Sheryl Gay Stolberg, and Noah Weiland. "Biden to
       Send 500 Million Doses of Pfizer Vaccine to 100 Countries Over a Year."
       New York Times, June 9, 2021; Updated June 30, 2021. https://www.
       nytimes.com/2021/06/09/us/politics/biden-Pfizer-vaccine-doses.html

116.   Stoddart, Michelle, and Sarah Kolinovsky. "Biden announces US
       to donate another half billion vaccine doses to lower-income
       nations." *ABC News*, September 22, 2021. https://abcnews.go.com/
       Politics/biden-announces-us-donate-half-billion-vaccine-doses/
       story?id=80167793

References for Chapter 4

1. LaVergne, Stephanie. "How many people get 'long COVID' – and who is most at risk?." The Conversation, February 17, 2021. https://theconversation.com/how-many-people-get-long-covid-and-who-is-most-at-risk-154331

2. WorldOMeter. "Coronavirus: United States." Accessed September 30, 2021. https://www.worldometers.info/coronavirus/country/us/

3. Liu, Yang and Joacim Roclöv. "The Reproductive Number of the Delta Variant of SARS-CoV-2 is Far Higher Compared to the Ancestral SARS-CoV-2 Virus." https://pubmed.ncbi.nlm.nih.gov/34369565/?

4. Armstrong, Drew. "The World's Most Loathed Industry Gave Us a Vaccine in Record Time." *Bloomberg Businessweek,* December 23, 2020. https://www.bloomberg.com/news/features/2020-12-23/covid-vaccine-how-big-pharma-saved-the-world-in-2020

5. Hargreaves, Eilidh. "Private Members Club Vaccinating Clients Abroad Is 'Proud' to Offer the Service." *Telegraph,* January 12, 2021. https://www.telegraph.co.uk/luxury/society/private-members-club-vaccinating-clients-abroad-proud-offer/

6. Carmichael, Flora, and Jack Goodman. "Vaccine Rumours Debunked: Microchips, 'Altered DNA' and More." *BBC News,* December 2, 2020. https://www.bbc.co.uk/news/54893437

7. Estrin, Daniel. "How Israel Persuaded Reluctant Ultra-Orthodox Jews To Get Vaccinated Against COVID-19." *NPR,* April 22, 2021. https://www.npr.org/2021/04/22/988812635/how-israel-persuaded-reluctant-ultra-Orthodox-jews-to-get-vaccinated-against-cov

8. Heller, Oren, Yaniv Shlomo, Yung Chun, Mary Acri, and Michal Grinstein-Weiss. "The Game Is Not Yet Over, and Vaccines Still Matter: Lessons From a Study on Israel's COVID-19 Vaccination." The Brookings Institute, September 13, 2021. https://www.brookings.edu/blog/up-front/2021/09/13/the-game-is-not-yet-over-and-vaccines-still-matter-lessons-from-a-study-on-israels-covid-19-vaccination/

9. Heard, Miriam Delaney. "Building Trust in the COVID Vaccine in Black Communities." National Health Law Program, February 24, 2021. https://healthlaw.org/building-trust-in-the-covid-vaccine-in-black-communities/

10. Author interview with Dr. Howard Heller, Infectious Disease Consultant at Harvard Medical School and Massachusetts General Hospital;

and Senior Advisor for Clinical Partnerships at the Massachusetts Consortium on Pathogen Readiness, on May 10, 2021.

11. Ibid.

12. Ibid.

13. Author interview with Gil Epstein, Professor of Economics and Dean of Social Science at Bar-Ilan University, April 18, 2021.

14. Wetsman, Nicole. "Vaccines Should be the New Bobbleheads at Every Sporting Event." The Verge, April 28, 2021. https://www.theverge. com/2021/4/28/22407753/covid-vaccine-milwaukee-bucks-arena-nba

15. Bravo, Christine. "COVID Shot While You Shop: California Pharmacies Now Offer Walk-Up Vaccine Appointments." NBC San Diego, May 5, 2021. https://www.nbcsandiego.com/news/local/covid-shot- while-you-shop-california-pharmacies-now-offer-walk-up-vaccine- appointments/2596418/

16. "UMD Part of New National Initiative that Taps Barbers and Stylists to Help Address Vaccine Hesitancy." University of Maryland School of Public Health, June 2, 2021. https://sph.umd.edu/news/umd-part-new- national-initiative-taps-barbers-and-stylists-help-address-vaccine- hesitancy

17. Sun, Lena H. "A New National Model? Barbershop Offers Coronavirus Shots in Addition to Cuts and Shaves." *Washington Post,* May 30, 2021. https://www.washingtonpost.com/health/2021/05/30/ barbershop-coronavirus-vaccines

18. Gold, Hadas. "Israel Vaccination 'Green Pass' May Offer a Glimpse of a Post-Covid Future." CNN, March 11, 2021. https://www.cnn.com/travel/ article/israel-vaccine-green-pass-wellness/index.html

19. Kershner, Isabel. "My Life in Israel's Brave New Post-Pandemic Future." *New York Times,* April 5, 2021. https://www.nytimes.com/2021/04/05/ world/middleeast/israel-vaccinations.html

20. Breeden, Aurelien. "France Approves a Contentious Law Making Health Passes Mandatory." *New York Times,* July 26, 2021. https://www.nytimes. com/2021/07/26/world/france-covid-pass-required.html

21. Howell, Beth. "Which Countries Are Using COVID-19 Vaccine Passports?" MoveHub, September 27, 2021. https://www.movehub.com/ blog/countries-using-covid-passports/

22. Caspani, Maria, and Dan Whitcomb. "New York Becomes First U.S. City to Order COVID Vaccines for Restaurants, Gyms."

Reuters, August 3, 2021. https://www.reuters.com/world/us/
nyc-require-proof-vaccination-indoor-activities-mayor-2021-08-03/

23.   Rouw, Anna, Jennifer Kates, and Josh Michaud. "Key Questions
      about COVID-19 Vaccine Passports and the U.S." Kaiser Family
      Foundation. https://www.kff.org/coronavirus-Covid-19/issue-brief/
      key-questions-about-Covid-19-vaccine-passports-and-the-u-s/

24.   Choi, Joseph. "Marjorie Taylor Greene Blasts COVID-19 Vaccine
      Passports: 'Biden's Mark of the Beast'." The Hill, March 30, 2021. https://
      thehill.com/homenews/house/545649-marjorie-taylor-greene-blasts-
      Covid-19-vaccine-passports-as-bidens-mark-of-the

25.   Colchester, Max, and Felicia Schwartz. "Covid-19 Vaccine 'Passports'
      Raise Ethics Concerns, Logistical Hurdles." *Wall Street Journal*, February
      26, 2021. https://www.wsj.com/articles/Covid-19-vaccine-passports-
      raise-ethics-concerns-logistical-hurdles-11614335403

26.   Beck, Luisa, and Loveday Morris. "Germany lets the vaccinated have a
      bit more fun. Teens shut out from the jabs are not happy." *Washington
      Post,* May 6, 2021. https://www.washingtonpost.com/world/europe/
      germany-coronavirus-vaccine-rules/2021/05/06/b6a17716-adbb-11eb-
      82c1-896aca955bb9_story.html

27.   Office of the Governor, West Virginia. "COVID-19 UPDATE: Gov. Justice:
      West Virginia offering $100 savings bond to residents age 16 to 35 who
      choose to get vaccinated." [Press release]. https://governor.wv.gov/
      News/press-releases/2021/Pages/COVID-19-UPDATE-Gov.-Justice-West-
      Virginia-offering-100-savings-bond-to-residents-age-16-to-35-who-get-
      vaccinated.aspx

28.   Vaughn, Hayley. "Guns, trucks and cash: West Virginia takes vaccine
      incentives to new levels." NBC News, June 2, 2021. https://www.
      nbcnews.com/news/us-news/guns-trucks-cash-west-virginia-takes-
      vaccine-incentives-new-levels-n1269352

29.   DeWine, Mike. "Don't Roll Your Eyes at Ohio's Vaccine Lottery." *New York
      Times,* May 26, 2021. https://www.nytimes.com/2021/05/26/opinion/
      ohio-vaccine-lottery-mike-dewine.html

30.   Mervosh, Sarah. "Meet the Ohio Vaccine Lottery's $1 Million Winner: A
      22-Year-Old Who 'Thought It Was a Prank.'" *New York Times,* May 26,
      2021. https://www.nytimes.com/2021/05/27/us/ohio-vaccine-lottery-
      winner.html

31.   DeWine, "Don't Roll Your Eyes at Ohio's Vaccine Lottery."

32.   Ibid.

33. Barber, Andrew, and Jeremy West. "Conditional Cash Lotteries Increase COVID-19 Vaccination Rates." (July 26, 2021). Available at SSRN: https://ssrn.com/abstract=3894034 or http://dx.doi.org/10.2139/ssrn.3894034

34. Shah, Saeed, and Waqar Gillani. "In Pakistan, Saying 'No' to Covid-19 Vaccine Carries Consequences." *Wall Street Journal*, June 22, 2021. https://www.wsj.com/articles/saying-no-to-Covid-19-vaccine-in-pakistan-carries-consequences-11624359601

35. Krispy Kreme Doughnut Corp. "Covid-19 Vaccine Offer." Accessed September 30, 2021. https://www.krispykreme.com/promos/vaccineoffer

36. Tyko, Kelly. "Budweiser giving away free beer for COVID vaccine with 'Reunited with Buds' giveaway. How to sign up." *USA Today*, April 15, 2021. https://www.usatoday.com/story/money/food/2021/04/15/budweiser-covid-vaccine-incentive-free-beer/7236622002/

37. Chappell, Bill. "Uber And Lyft Will Give Free Rides To COVID-19 Vaccination Spots, White House Says." *NPR*, May 11, 2021. https://www.npr.org/sections/coronavirus-live-updates/2021/05/11/995882805/uber-and-lyft-will-give-free-rides-to-covid-19-vaccination-spots-white-house-say

38. Aten, Jason. "United Airlines Is Giving Away a Year of Free Flights. There's Only 1 Catch." *Inc.*, May 26, 2021. https://www.inc.com/jason-aten/united-airlines-is-giving-away-a-year-of-free-flights-theres-only-1-catch.html

39. Dollar General Corporation. "Dollar General Removes Barriers for Frontline Workers to Get COVID-19 Vaccine." [Press release]. January 13, 2021. https://newscenter.dollargeneral.com/Covid-19/dollar-general-removes-barriers-for-frontline-workers-to-get-vaccine.htm

40. Ansari, Maira. "Kroger offering $100 to employees who get COVID vaccine." WAVE 3 News, February 6, 2021. https://www.wave3.com/2021/02/06/kroger-offering-employees-who-get-covid-vaccine/

41. Bellon, Tina, and Richa Naidu. "As U.S. companies push to get workers vaccinated, states disagree on who's essential." Reuters, December 8, 2020. https://www.reuters.com/article/us-health-coronavirus-vaccine-patchwork/as-u-s-companies-push-to-get-workers-vaccinated-states-disagree-on-whos-essential-idUSKBN28I2OD

42. Friedman, Gillian, and Lauren Hirsch. "Help With Vaccination Push Comes From Unexpected Businesses." *New York Times*, May 7, 2021. https://www.nytimes.com/2021/01/23/business/vaccines-microsoft-amazon-starbucks.html

43. Andrew, Scottie. "A Texas Hospital System Will Require Employees to Get the Covid-19 Vaccine and Could Fire Them If They Don't Comply." *CNN*, April 26, 2021. https://www.cnn.com/2021/04/26/us/houston-methodist-covid-vaccine-mandate-trnd/index.html

44. Dickler, Jessica. "Hundreds of Colleges Say Covid Vaccines Will Be Mandatory for Fall 2021." *CNBC*, May 11, 2021. https://www.cnbc.com/2021/05/11/hundreds-of-colleges-to-require-covid-vaccines-for-fall-2021.html

45. Schmidt, Martin A. "Requiring Employees to be Vaccinated, and Other Updates." MIT Letter to the Community. June 1, 2021. http://orgchart.mit.edu/node/6/letters_to_community/requiring-employees-be-vaccinated-and-other-updates

46. Office of the President, Yale University. "Requiring Faculty, Staff, and Trainees to be Vaccinated Against COVID-19." May 14, 2021. https://president.yale.edu/president/statements/requiring-faculty-staff-and-trainees-be-vaccinated-against-Covid-19

47. Thomason, Andy, and Brian O'Leary. "Here's a List of Colleges That Require Students or Employees to Be Vaccinated Against Covid-19." *Chronicle of Higher Education,* September 28, 2021. Updated October 1, 2021. https://www.chronicle.com/blogs/live-coronavirus-updates/heres-a-list-of-colleges-that-will-require-students-to-be-vaccinated-against-Covid-19

48. Waddell, Melanie. "Wall Street Vaccination Policies: What 9 Firms Are Requiring." [slideshow]. Think Advisor, August 31, 2021. https://www.thinkadvisor.com/2021/08/31/wall-street-vaccination-policies-what-9-firms-are-requiring/

49. Rogers, Katie, and Sheryl Gay Stolberg. "Biden Mandates Vaccines for Workers, Saying, 'Our Patience Is Wearing Thin'." *New York Times*, September 9, 2021. Updated September 22, 2021. https://www.nytimes.com/2021/09/09/us/politics/biden-mandates-vaccines.html

50. Goldstein, Amy. "Should Health-Care Workers Be Required to Get Coronavirus Shots? Companies Grapple with Mandates." *Washington Post,* April 5, 2021. https://www.washingtonpost.com/health/should-health-care-workers-be-required-to-get-coronavirus-shots-companies-grapple-with-mandates/2021/04/04/2369048e-92f2-11eb-a74e-1f4cf89fd948_story.html

51. U.S. Equal Employment Opportunity Commission. "What You Should Know About COVID-19 and the ADA, the Rehabilitation Act, and Other EEO Laws." Department of Labor. https://www.eeoc.gov/wysk/

what-you-should-know-about-Covid-19-and-ada-rehabilitation-act-and-other-eeo-laws

52. Norwegian Cruise Line. "Sail Safe." Accessed October 1, 2021. https://www.ncl.com/sail-safe

53. Office of the Governor, State of Florida. "Executive Order Number 21-91 (Prohibiting COVID-19 Vaccine Passports)." Accessed October 1, 2021. https://www.flgov.com/wp-content/uploads/2021/04/EO-21-81.pdf

54. Fuller, Austin. "Florida COVID 'vaccine passport' fines begin today for businesses, governments." *Orlando Sentinel*, September 15, 2021. https://www.orlandosentinel.com/business/os-bz-vaccine-passport-law-florida-20210915-ul2id2egibczpmwujcaybczose-story.html

55. Kallingal, Mallika and Andy Rose. "Major cruise ship company may avoid Florida if state doesn't permit Covid-19 vaccination checks." *CNN*, May 27, 2021. https://www.cnn.com/travel/article/norwegian-cruise-line-florida-vaccinations/index.html

56. Chokshi, Niraj. "Norwegian Cruise Line Holdings Sues Florida Over Prohibition on Vaccine Requirements." *New York Times*, July 13, 2021. Updated August 9, 2021. https://www.nytimes.com/2021/07/13/business/norwegian-cruise-florida-vaccine-requirement.html

57. "Federal Judge Sides with Norwegian Cruise Line in Fight with Florida Over Vaccine Passports." *Politico*, August 8, 2021. https://www.politico.com/states/florida/story/2021/08/08/federal-judge-sides-with-norwegian-cruise-line-in-fight-with-florida-over-vaccine-passports-1389732

58. Fox, Alison. "Every Cruise Line Requiring Passengers to Be Vaccinated Before Boarding." *Travel + Leisure*, August 25, 2021. https://www.travelandleisure.com/cruises/cruises-that-allow-vaccinated-travelers

59. Associated Press. "Florida to Start Enforcing $5,000 Fine for Seeking Proof of COVID Vaccine." *Orlando Sentinel*, September 2, 2021. https://www.orlandosentinel.com/business/os-bz-vaccine-passports-florida-fine-20210902-h4yu6svpi5atzeot6jupnm4op4-story.html

60. Husch Blackwell. "50-state Update on Pending Legislation Pertaining to Employer-mandated Vaccinations." March 5, 2021. Updated July 1, 2021. https://www.huschblackwell.com/newsandinsights/50-state-update-on-pending-legislation-pertaining-to-employer-mandated-vaccinations

61. "Mandatory Vaccination Policy Lawsuit Update: Nurses Take a Shot Against Hospital, But Judge Jabs Back." *National Law Review*, June 15, 2021. https://www.natlawreview.com/article/

mandatory-vaccination-policy-lawsuit-update-nurses-take-shot-against-hospital-judge

62. Russo, Salvatore. "Houston Methodist Hospital Loses 153 Employees Over Mandatory COVID-19 Vaccination Policy." JD Supra, June 30, 2021. https://www.jdsupra.com/legalnews/houston-methodist-hospital-loses-153-8268931/

63. Murphy Marcos, Coral. "United Airlines to Fire Workers Who Refused to Get a Vaccination." *New York Times*, September 29, 2021. https://www.nytimes.com/2021/09/29/business/united-airlines-vaccine-mandate.html

64. Food & Drug Administration. "Joint CDC and FDA Statement on Johnson & Johnson COVID-19 Vaccine." [Press release]. April 13, 2021. https://www.fda.gov/news-events/press-announcements/joint-cdc-and-fda-statement-johnson-johnson-Covid-19-vaccine

65. Nordquist, Richard. "What Is a Post Hoc Logical Fallacy?" ThoughtCo., January 17, 2020. https://www.thoughtco.com/post-hoc-fallacy-1691650

66. Shimabukuro, Tom, for the Advisory Committee on Immunization Practices. "Thrombosis with Thrombocytopenia Syndrome (TTS) Following Janssen COVID-19 Vaccine." [slideshow]. Centers for Disease Control and Prevention, National Center for Immunization & Respiratory Diseases. April 23, 2021. https://www.cdc.gov/vaccines/acip/meetings/downloads/slides-2021-04-23/03-COVID-Shimabukuro-508.pdf

67. Ibid.

68. WorldOMeter. "Coronavirus: United States." Accessed April 24, 2021.

69. Vazquez, Jeanna. "Study: COVID-19 Infection Combined with Blood Clots Worsen Patient Outcomes." [Press release]. UC San Diego Health, November 23, 2020. https://health.ucsd.edu/news/releases/Pages/2020-11-23-study-Covid-19-infection-combined-with-blood-clots-worsen-patient-outcomes.aspx

70. Saey, Tina Hesman. "FDA and CDC OK Resuming J&J COVID-19 Shots Paused Over Rare Clot Concerns." *Science News*, April 23, 2021. https://www.sciencenews.org/article/johnson-janssen-covid-coronavirus-vaccine-resume-pause-rare-blood-clot

71. Drake, John. "The Risk Of Covid-19 Is Greater Than Blood Clots From The Johnson & Johnson Vaccine." *Forbes*, April 30, 2021. https://www.forbes.com/sites/johndrake/2021/04/30/

the-risk-of-covid-19-is-far-greater-than-blood-clots-from-the-
johnson—johnson-vaccine/?sh=6c472f9b28a3

72.   Kupferschmidt, Kai, and Gretchen Vogel. "Hard Choices Emerge
       as Link Between AstraZeneca Vaccine and Rare Clotting Disorder
       Becomes Clearer." *Science*, April 11, 2021. https://www.sciencemag.org/
       news/2021/04/hard-choices-emerge-link-between-AstraZeneca-vaccine-
       and-rare-clotting-disorder-becomes

73.   Mueller, Benjamin. "AstraZeneca Vaccine Faces New Setbacks in U.K. and
       European Union." *New York Times*, April 7, 2021. https://www.nytimes.
       com/2021/04/07/world/europe/AstraZeneca-uk-european-union.html

74.   "Germany restricts use of AstraZeneca vaccine to over 60s in most
       cases." Deutsche Welle. Accessed October 1, 2021. https://www.dw.com/
       en/germany-restricts-use-of-AstraZeneca-vaccine-to-over-60s-in-most-
       cases/a-57049301

75.   Department of Health, Australian Government. "Vaxzevria
       (AstraZeneca)." Accessed October 1, 2021. https://www.
       health.gov.au/initiatives-and-programs/Covid-19-vaccines/
       learn-about-Covid-19-vaccines/about-the-AstraZeneca-Covid-19-vaccine

76.   "J&J Vaccine: Side Effects, Guillain-Barre Syndrome and What You
       Should Know." NBC Chicago, July 13, 2021. https://www.nbcchicago.com/
       news/coronavirus/jj-vaccine-side-effects-guillain-barre-syndrome-and-
       what-you-should-know/2553296/

77.   Pollard, James. "What to Know About Guillain-
       Barré Syndrome." NBC Chicago, July 12, 2021. https://
       www.nbcchicago.com/news/national-international/
       what-to-know-about-guillain-barre-syndrome/2552690/

78.   Centers for Disease Control and Prevention. "General Questions and
       Answers on Guillain-Barré syndrome (GBS)." December 15, 2009. https://
       www.cdc.gov/h1n1flu/vaccination/gbs_qa.htm

79.   Diaz, George A., Guilford T. Parsons, Sara K. Gering, et al. "Myocarditis
       and Pericarditis After Vaccination for COVID-19." *JAMA*, 326 no.
       12 (2021): 1210–1212. https://jamanetwork.com/journals/jama/
       fullarticle/2782900

80.   Centers for Disease Control and Prevention. "Investigating Long-
       Term Effects of Myocarditis." August 20, 2021. https://www.cdc.
       gov/coronavirus/2019-ncov/vaccines/safety/myo-outcomes.html

81.   Vogel, Gretchen, and Jennifer Couzin-Frankel. "Israel Reports Link
       Between Rare Cases of Heart Inflammation and COVID-19 Vaccination

in Young Men." *Science,* June 1, 2021. https://www.sciencemag.org/news/2021/06/israel-reports-link-between-rare-cases-heart-inflammation-and-Covid-19-vaccination

82. Boehmer, Tegan K., Lyudmyla Kompaniyets, Amy M. Lavery, et al. "Association Between COVID-19 and Myocarditis Using Hospital-Based Administrative Data — United States, March 2020–January 2021." *Morbidity and Mortality Weekly Report (MMWR)* 70 no. 35 (2021): 1228–32. https://www.cdc.gov/mmwr/volumes/70/wr/mm7035e5.htm

83. Centers for Disease Control and Prevention. "About Dengue: What You Need to Know." Page last reviewed September 23, 2021. https://www.cdc.gov/dengue/about/index.html

84. Uildriks, Lori. "COVID-19's Impact on Dengue Transmission." *Medical News Today,* November 4, 2020. https://www.medicalnewstoday.com/articles/covid-19s-impact-on-dengue-transmission

85. Lo, Chris. "Dengue Vaccine Dilemma." *Pharmaceutical Technology,* December 15, 2019. https://www.pharmaceutical-technology.com/features/dangvaxia-philippines/

86. Ibid.

87. Author interview with Dr. Howard Heller, May 10, 2021.

88. WHO data cited in Soucheray, Stephanie. "Philippines, scientists grapple with Dengvaxia fallout." Center for Infectious Disease Research and Policy (CIDRAP), December 14, 2017. https://www.cidrap.umn.edu/news-perspective/2017/12/philippines-scientists-grapple-dengvaxia-fallout

89. Ibid.

90. Author interview with Dr. Howard Heller, May 10, 2021.

91. Beaubien, Jason. "The Philippines Is Fighting One Of The World's Worst Measles Outbreaks." *NPR,* May 23, 2019. https://www.npr.org/sections/goatsandsoda/2019/05/23/725726094/the-philippines-is-fighting-one-of-the-worlds-worst-measles-outbreaks

92. Young, Saundra. "Black Vaccine Hesitancy Rooted in Mistrust, Doubts." WebMD, February 2, 2021. https://www.webmd.com/vaccines/Covid-19-vaccine/news/20210202/black-vaccine-hesitancy-rooted-in-mistrust-doubts

93. Elliott, Debbie. "In Tuskegee, Painful History Shadows Efforts To Vaccinate African Americans." *NPR,* February 16, 2021. https://www.npr.org/2021/02/16/967011614/

in-tuskegee-painful-history-shadows-efforts-to-vaccinate-african-americans

94.   "Henrietta Lacks: science must right a historical wrong." *Nature*, 585 no. 7823, (2020): 7. https://www.nature.com/articles/d41586-020-02494-z

95.   Khazan, Olga. "The Tucker Carlson Fans Who Got Vaxxed." *Atlantic*, August 9, 2021. https://www.theatlantic.com/politics/archive/2021/08/why-so-many-republicans-wont-get-vaccinated/619659/

96.   Ivory, Danielle, Lauren Leatherby, and Robert Gebeloff. "Least Vaccinated U.S. Counties Have Something in Common: Trump Voters." *New York Times*, April 17, 2021. https://www.nytimes.com/interactive/2021/04/17/us/vaccine-hesitancy-politics.html

97.   White, Marcus. "Here's Where – and Why – San Francisco Marathon Runners Will Need to Wear Masks." *MSN*, September 17, 2021. https://www.msn.com/en-us/travel/news/heres-where-and-why-san-francisco-marathon-runners-will-need-to-wear-masks/ar-AAOyxYf

98.   Kahneman, Daniel. "Thinking Fast and Slow," Toronto, ON: Doubleday Canada, 2011.

99.   Baer, Drake. "Kahneman: Your Cognitive Biases Act Like Optical Illusions." *The Cut*, New York Magazine, January 13, 2017. https://www.thecut.com/2017/01/kahneman-biases-act-like-optical-illusions.html

100. Frankovic, Kathy. "One in Five Americans Continue to Refuse the COVID-19 Vaccine." YouGovAmerica, May 6, 2021. https://today.yougov.com/topics/politics/articles-reports/2021/05/06/one-five-americans-continue-refuse-covid-19-vaccin

101. Wiggins, Ovetta. "Maryland Gave Away $2 Million in a Vaccine Lottery to Boost Vaccinations. Did It Work?" *Washington Post*, August 1, 2021. https://www.washingtonpost.com/local/covid-vaccine-lottery-maryland/2021/07/30/0081d91e-eb0d-11eb-8950-d73b3e93ff7f_story.html

102. Kahneman, "Thinking fast and Slow."

103. Kaplan, Robert M. "A False Narrative About 'Misinformation' and Covid Vaccines." *Wall Street Journal*, August 3, 2021. https://www.wsj.com/articles/misinformation-covid-vaccine-hesitancy-polling-demographics-polarization-partisanship-11627914431?mod=opinion_lead_pos9

104. West, Melanie Grayce, and Talal Ansari. "Delta Variant Fuels Missouri's Covid-19 Uptick." *Wall Street Journal*, July 3, 2021. https://www.wsj.com/articles/delta-variant-fuels-missouris-Covid-19-uptick-11625304601

105. Beaubien, Jason. "A Cow Head Will Not Erupt From Your Body If You Get A Smallpox Vaccine." *NPR*, January 7, 2015. https://www.npr.org/sections/goatsandsoda/2015/01/07/375598652/a-cow-head-will-not-erupt-from-your-body-if-you-get-a-smallpox-vaccine

106. "History of Anti-vaccination Movements." The History of Vaccines. Accessed October 1, 2021. https://www.historyofvaccines.org/content/articles/history-anti-vaccination-movements

107. Bremner, Charles. "One Third of French Say Vaccines Are Dangerous." *Times* (London), June 20, 2019. https://www.thetimes.co.uk/article/one-third-of-french-say-vaccines-are-dangerous-22cvm9nwg

108. Pesce, Nicole Lyn. "This Is the Most Anti-Vaxxer Country in the World." *MarketWatch*, June 19, 2019. https://www.marketwatch.com/story/this-is-the-most-anti-vaxxer-country-in-the-world-2019-06-19

109. Kershner, Isabel. "Israel's Early Vaccine Data Offers Hope." *New York Times*, January 25, 2021. https://www.nytimes.com/2021/01/25/world/middleeast/israels-vaccine-data.html

110. Fiordaliso, Michelle. "I'm Unvaccinated. It's Not What You Think. *Washington Post*, July 17, 2021. https://www.washingtonpost.com/health/covid-unvaccinated-medical-reason/2021/07/16/4937475a-df3e-11eb-9f54-7eee10b5fcd2_story.html

111. Center for Countering Digital Hate. "The Disinformation Dozen: Why Platforms Must Act on Twelve Leading Online Anti-Vaxxers." CCDH, March 24, 2021. https://www.counterhate.com/disinformationdozen

112. Office of the Attorney General, State of Connecticut. "Vaccine Disinformation." [Letter]. March 24, 2021. https://portal.ct.gov/-/media/AG/Press_Releases/2021/AG-Letter-to-Tech-CEOs.pdf

113. Szaniszlo, Marie. "AGs call on Facebook, Twitter to stop spread of 'deadly' coronavirus vaccine 'disinformation'." *Boston Herald*, March 24, 2021. https://www.bostonherald.com/2021/03/24/ags-call-on-facebook-twitter-to-stop-spread-of-deadly-coronavirus-vaccine-disinformation/

114. Guynn, Jessica. "Facebook and Twitter Must Crack Down On COVID-19 Vaccine Hoaxes and Lies, 12 State Attorneys General say." *USA Today*, March 24, 2021. https://www.usatoday.com/story/tech/2021/03/24/facebook-twitter-covid-vaccine-lies-state-attorneys-general/6983079002/

115. "The Facebook Files: A Wall Street Journal Investigation." *Wall Street Journal*. Accessed October 1, 2021. https://www.wsj.com/articles/the-facebook-files-11631713039

116. Schechner, Sam, Jeff Horwitz, and Emily Glazer. "How Facebook Hobbled Mark Zuckerberg's Bid to Get America Vaccinated." *Wall Street Journal*, September 17, 2021. https://www.wsj.com/articles/ facebook-mark-zuckerberg-vaccinated-11631880296?mod=article_inline

117. Ortutay, Barbara, and Amanda Seitz. "Defying Rules, Anti-Vaccine Accounts Thrive on Social Media." *ABC News*, March 12, 2021. https://abcnews.go.com/Technology/wireStory/ defying-rules-anti-vaccine-accounts-thrive-social-media-76410129

118. Benton, Joshua. "What Sort of News Travels Fastest Online? Bad News, You Won't Be Shocked to Hear." NiemanLab, July 15, 2019. https://www. niemanlab.org/2019/07/what-sort-of-news-travels-fastest-online-bad-news-you-wont-be-shocked-to-hear/

119. Dizikes, Peter. Study: On Twitter, False News Travels Faster Than True Stories. *MIT News*, March 8, 2018. https://news.mit.edu/2018/ study-twitter-false-news-travels-faster-true-stories-0308

120. Ward, Alex. "World Leaders Who Denied the Coronavirus's Danger Made Us All Less safe." *Vox*, March 30, 2020. https://www.vox.com/2020/3/30 /21195469/coronavirus-usa-china-brazil-mexico-spain-italy-iran

121. "Coronavirus: Iran Cover-Up of Deaths Revealed by Data Leak." *BBC News*, August 3, 2020. https://www.bbc.com/news/ world-middle-east-53598965

122. Ward, Alex. "Mexico's Coronavirus-Skeptical President Is Setting Up His Country for a Health Crisis." *Vox*, March 26, 2020. Updated March 28, 2020. https://www.vox.com/2020/3/26/21193823/coronavi rus-mexico-andres-manuel-lopez-obrador-health-care

123. McCoy, Terrence, and Heloísa Traiano. "Brazil's Bolsonaro, Channeling Trump, Dismisses Coronavirus Measures — It's Just 'a Little Cold'." *Washington Post*, March 25, 2020. https://www.washingtonpost.com/ world/the_americas/brazils-bolsonaro-channeling-trump-dismisses-coronavirus-measures—its-just-a-little-cold/2020/03/25/65bc90d6-6e99-11ea-a156-0048b62cdb51_story.html

124. Montanaro, Domenico. "FACT CHECK: Trump Compares Coronavirus To The Flu, But It Could Be 10 Times Deadlier." *NPR*, March 24, 2020. https://www.npr.org/sections/coronavirus-live-updates/2020/03/24/820797301/fact-check-trump-compares-coronavirus-to-the-flu-but-they-are-not-the-same

125. Minder, Raphael. "Spain Becomes Latest Epicenter of Coronavirus After a Faltering Response." *New York Times*, March 13, 2020. https://www.

nytimes.com/2020/03/13/world/europe/spain-coronavirus-emergency.
html

126. Ward, Philip. "What Can You Do to Prevent Clinical
      Trial Fraud?" *Applied Clinical Trials,* December 1, 2017.
      https://www.appliedclinicaltrialsonline.com/view/
      what-can-you-do-prevent-clinical-trial-fraud

127. Fanelli, Daniele. "How many scientists fabricate and falsify research? A
      systematic review and meta-analysis of survey data." *PLoS One,* 4 no. 5
      (2009): e5738. https://pubmed.ncbi.nlm.nih.gov/19478950/

128. Baumgartner, Frank R., and Bryan D. Jones. "Agenda Dynamics and
      Policy Subsystems." *Journal of Politics,* 53 no. 4 (November 1991): 1044–74.
      https://doi.org/10.2307/2131866.

129. Lopez, German. "The Reagan Administration's Unbelievable Response
      to the HIV/AIDS Epidemic." *Vox,* December 1, 2016. https://www.vox.
      com/2015/12/1/9828348/ronald-reagan-hiv-aids

130. Bush, George W. "George W. Bush: PEPFAR saves millions of lives in
      Africa. Keep it fully funded." *Washington Post,* April 7, 2017. https://www.
      washingtonpost.com/opinions/george-w-bush-pepfar-saves-millions-
      of-lives-in-africa-keep-it-fully-funded/2017/04/07/2089fa46-1ba7-11e7-
      9887-1a5314b56a08_story.html

131. Kessler, Glenn, Adrian Blanco, and Tyler Remmel. "The False and
      Misleading Claims President Biden Made During His First 100
      Days in Office." *Washington Post,* April 25, 2021. Updated April 30,
      2021. https://www.washingtonpost.com/politics/interactive/2021/
      biden-fact-checker-100-days/

132. Gordon, Michael R., and Dustin Volz. "Russian Disinformation
      Campaign Aims to Undermine Confidence in Pfizer, Other Covid-19
      Vaccines, U.S. Officials Say." *Wall Street Journal,* March 7, 2021. https://
      www.wsj.com/articles/russian-disinformation-campaign-aims-to-
      undermine-confidence-in-Pfizer-other-Covid-19-vaccines-u-s-officials-
      say-11615129200

133. Glaun, Dan. "Chinese Spam Network Aims to Discredit U.S. COVID
      Vaccine and Response, Report Finds." *Frontline,* February 4, 2021.
      https://www.pbs.org/wgbh/frontline/article/chinese-spam-network-
      aims-to-discredit-u-s-covid-vaccine-and-response-report-finds/

134. "EEAS Special Report Update: Short Assessment of Narratives and
      Disinformation Around the COVID-19 Pandemic (Update December
      2020 - April 2021)." EUvsDisinfo.eu, April 28, 2021. https://euvsdisinfo.
      eu/eeas-special-report-update-short-assessment-of-narratives-and-

disinformation-around-the-Covid-19-pandemic-update-december-
2020-april-2021/

135. Ibid.

136. Frenkel, Sheera, Maria Abi-Habib, and Julian E. Barnes. "Russian
Campaign Promotes Homegrown Vaccine and Undercuts Rivals."
New York Times, February 5, 2021. Updated February 6, 2021. https://
www.nytimes.com/2021/02/05/technology/russia-covid-vaccine-
disinformation.html

137. Edelman. "2021 Edelman Trust Barometer." Daniel J. Edelman
Holdings, Inc. Accessed October 1, 2021. https://www.edelman.com/
trust/2021-trust-barometer

138. Haelle, Tara. "Vaccine Hesitancy Is Nothing New. Here's
the Damage It's Done Over Centuries." *Science News,*
May 11, 2021. https://www.sciencenews.org/article/
vaccine-hesitancy-history-damage-anti-vaccination

139. Cohen, Elizabeth. "Fauci Says Covid-19 Vaccine May Not Get US to Herd
Immunity If Too Many People Refuse to Get It." *CNN,* June 28, 2020.
https://www.cnn.com/2020/06/28/health/fauci-coronavirus-vaccine-
contact-tracing-aspen/index.html

140. Biden, Joe. Twitter post, October 28, 2020, 8:15 PM. https://twitter.com/
JoeBiden/status/1321606423495823361

141. Varadarajan, Tunku. "How Science Lost the Public's Trust." *Wall Street
Journal,* July 23, 2021. https://www.wsj.com/articles/covid-china-media-
lab-leak-climate-ridley-biden-censorship-coronavirus-11627049477

142. Azoulay, Pierre, Christian Fons-Rosen, and Joshua S. GraffZivin.
"Does Science Advance One Funeral at a Time?" *American Economic
Review* 109 no. 8 (August 2019): 2889–2920. https://www.aeaweb.org/
articles?id=10.1257/aer.20161574

143. Estrin, "How Israel Persuaded Reluctant Ultra-Orthodox Jews To Get
Vaccinated Against COVID-19."

144. Ibid.

145. Ibid.

146. Scharf, Isaac, and Ilan Ben Zion. "As Vaccinations Lag, Israel Combats
Online Misinformation." *AP News,* February 15, 2021. https://
apnews.com/article/israel-misinformation-coronavirus-pandemic-
local-governments-museums-462b98e0ff0d7d5bf4400c84a610
47b2

147. Estrin, "How Israel Persuaded Reluctant Ultra-Orthodox Jews To Get Vaccinated Against COVID-19."

148. "How Do We Know the COVID-19 Vaccine Won't Have Long-Term Side Effects?" MU Healthcare. Accessed October 1, 2021. https://www.muhealth.org/our-stories/how-do-we-know-Covid-19-vaccine-wont-have-long-term-side-effects

149. Kostov, Nick. "How France Overcame Covid-19 Vaccine Hesitancy." *Wall Street Journal,* September 27, 2021. https://www.wsj.com/articles/how-france-overcame-covid-19-vaccine-hesitancy-11632735002

150. "The Right and Wrong Ways to Reduce Vaccine Hesitancy." *Economist,* July 24, 2021. https://www.economist.com/graphic-detail/2021/07/24/the-right-and-wrong-ways-to-reduce-vaccine-hesitancy

151. Ibid.

152. WorldOMeter. "Coronavirus: Singapore." Accessed October 1, 2021. https://www.worldometers.info/coronavirus/country/singapore/

References for Chapter 5

1. Jones, David and Stefan Helmreich. "A history of herd immunity." *Lancet*, September 19, 2020. https://www.thelancet.com/journals/lancet/article/PIIS0140-67362031924-3/fulltext

2. "What You Need to Know About Being Immunocompromised During COVID-19." Penn Medicine Health and Wellness [blog]. May 13, 2020. https://www.pennmedicine.org/updates/blogs/health-and-wellness/2020/may/what-it-means-to-be-immunocompromised

3. Mitnick, Joshua, and Antonio Regalado. "A Leaked Report Shows Pfizer's Vaccine Is Conquering Covid-19 in Its Largest Real-World Test." *MIT Technology Review*, February 19, 2021. https://www.technologyreview.com/2021/02/19/1019264/a-leaked-report-Pfizers-vaccine-conquering-Covid-19-in-its-largest-real-world-test/

4. Bendix, Aria. "Israel Offers a Glimpse of Life After Herd Immunity: With 80% of Adults Vaccinated, Cases Have Dropped to 15 Per Day." *Business Insider*, June 1, 2021. https://www.businessinsider.com/israel-vaccinated-most-adults-covid-herd-immunity-2021-6

5. Centers for Disease Control and Prevention. "SARS-CoV-2 Variant Classifications and Definitions." Updated September 23, 2021. https://www.cdc.gov/coronavirus/2019-ncov/variants/variant-info.html

6. Mendez, Rich. "Delta Variant is One of the Most Infectious Respiratory Diseases Known, CDC Director Says." *CNBC*, July 22, 2021. https://www.cnbc.com/2021/07/22/delta-variant-is-one-of-the-most-infectious-respiratory-diseases-known-cdc-director-says-.html

7. Schneider, Mike. "Florida breaks record for COVID-19 hospitalizations." *AP News*, August 1, 2021. https://apnews.com/article/business-health-florida-coronavirus-pandemic-7ca97f0d685ab25559cf9b51cfc077eb

8. WorldOMeter. "Coronavirus: Israel." Accessed August 3, 2021. https://www.worldometers.info/coronavirus/country/israel/

9. Schuster-Bruce, Catherine. "4 Coronavirus Variants Can Make People Sicker or Spread Faster, Including the Delta Variant First Found in India. Here's What Variants Are, and Why Experts Are So Concerned About Them." *Business Insider*, June 9, 2021. https://www.businessinsider.com/coronavirus-variants-uk-south-africa-brazil-us-facts-questions-2021-1

10.   Maxmen, Amy. "One Million Coronavirus Sequences: Popular Genome Site Hits Mega Milestone." *Nature* 593 (2021): 21. https://www.nature.com/articles/d41586-021-01069-w

11.   Nextstrain. "Genomic Epidemiology of Novel Coronavirus - Global Subsampling." Accessed October 2, 2021. https://nextstrain.org/ncov/global

12.   Centers for Disease Control and Prevention. "SARS-CoV-2 Variant Classifications and Definitions."

13.   See Table 3 on p. 26 of Food & Drug Administration. "Fact Sheet for Healthcare Providers Emergency Use Authorization (EUA) of Bamlanivimab and Etesevimab." Accessed October 2, 2021. https://www.fda.gov/media/145802/download

14.   Yan, Holly. "More Young People Are Getting Hospitalized as a 'Stickier,' More Infectious Coronavirus Strain Becomes Dominant." *CNN*, April 18, 2021. https://www.cnn.com/2021/04/12/health/b117-covid-variant-young-patients/index.html

15.   "Countries That Curbed Covid-19 Fast Have Been Slow to Vaccinate." *Economist,* March 6, 2021. https://www.economist.com/asia/2021/03/06/countries-that-curbed-Covid-19-fast-have-been-slow-to-vaccinate

16.   WorldOMeter. "Coronavirus: Australia." Accessed August 14, 2021. https://www.worldometers.info/coronavirus/country/australia/

17.   Cha, Sangmi. "S. Korea's Vaccination Drive Picks Up Speed, Little Slow Down in New Infections." Reuters, June 8, 2021. https://www.reuters.com/world/asia-pacific/skoreas-vaccination-drive-picks-up-speed-little-slow-down-new-infections-2021-06-08/

18.   Acro Biosystems. "Emerging Mutants from SARS-CoV-2 Variants." Accessed October 2, 2021. https://www.acrobiosystems.com/A1226-SARS-CoV-2_spike_mutants.html?gclid=CjwKCAjwzruGBhBAEiwAUqMR8IschX9_Dk1oLK1PpJkLFTpfx5n3oxdw6lIsfxsYzTJf-Y-cOvYcvxoClSoQAvD_BwE

19.   Lee, Jack J. "Coronavirus Mutations Could Muddle COVID-19 PCR Tests." *The Scientist,* May 17, 2021. https://www.the-scientist.com/news-opinion/coronavirus-mutations-could-muddle-Covid-19-pcr-tests-68772

20.   "Britain Reports Steep Rise in Weekly Delta Variant Cases." Reuters, June 18, 2021. https://www.reuters.com/world/uk/britain-says-33630-new-cases-delta-coronavirus-variant-latest-week-2021-06-18/

21.   Lin, Rong-Gong II, Luke Money, and Alex Wiggleworth. "Highly Contagious Delta Coronavirus Variant Spreading

Fast in California." *Los Angeles Times,* June 27, 2021.
https://www.latimes.com/california/story/2021-06-27/
highly-contagious-delta-coronavirus-variant-spreading-in-california

22.   GlobalData Healthcare. "Covid-19 Vaccine Effectiveness
      Affected by Variants." *Pharmaceutical Technology,* March 3,
      2021. https://www.pharmaceutical-technology.com/comment/
      Covid-19-vaccine-effectiveness-affected-by-variants/

23.   Collins, Francis. "Tracking the Evolution of a 'Variant
      of Concern' in Brazil." NIH Director's Blog, April
      27, 2021. https://directorsblog.nih.gov/2021/04/27/
      tracking-the-evolution-of-a-variant-of-concern-in-brazil/

24.   Booth, William, and Carolyn Y. Johnson. "South Africa suspends
      Oxford-AstraZeneca vaccine rollout after researchers report 'minimal'
      protection against coronavirus variant." *Washington Post,* February 7,
      2021. https://www.washingtonpost.com/world/europe/AstraZeneca-
      oxford-vaccine-south-african-variant/2021/02/07/e82127f8-6948-11eb-
      a66e-e27046e9e898_story.html

25.   Armour, Stephanie, and Jared S. Hopkins. "FDA Covid-19 Vaccine Booster
      Plan Could Be Ready Within Weeks." *Wall Street Journal,* August 5, 2021.
      https://www.wsj.com/articles/fda-Covid-19-vaccine-booster-plan-could-
      be-ready-within-weeks-11628194767

26.   Rabinovitch, Ari. "Israel offers COVID-19 booster to all vaccinated
      people." Reuters, August 29, 2021. https://www.reuters.com/world/
      middle-east/israel-offers-Pfizer-Covid-19-vaccine-booster-shots-adults-
      risk-2021-07-11/

27.   Beer, Tommy. "Israel To Offer Covid Vaccine Booster Shots To People
      Over 60." *Forbes,* July 29, 2021. https://www.forbes.com/sites/
      tommybeer/2021/07/29/israel-to-offer-covid-vaccine-booster-shots-to-
      people-over-60/?sh=734886246ec2

28.   Black, Derek. "Third Installment for All Citizens Over the Age of 12
      in Israel." World Stock Market. https://www.worldstockmarket.net/
      third-installment-for-all-citizens-over-the-age-of-12-in-israel/

29.   Smout, Alistair, and Young, Sarah. "COVID-19 booster
      vaccine campaign begins in England." Reuters,
      September 16, 2021. https://www.reuters.com/world/uk/
      covid-19-booster-vaccine-campaign-begins-england-2021-09-16/

30.   York, Joanna. "Covid France: Who Should Get a Third
      Booster Vaccination in Autumn?" *Connexion,* July 13,

2021. https://www.connexionfrance.com/French-news/
Covid-France-Who-should-get-a-third-booster-vaccination-in-autumn

31.   Bennhold, Katrin. "Germany Will Offer Vaccine Booster Shots Starting in September." *New York Times*, August 2, 2021. https://www.nytimes.com/2021/08/02/world/europe/coronavirus-booster-shots-germany.html

32.   Rall, Ted. "Why I Got a Third Covid Shot." *Wall Street Journal*, August 4, 2021. https://www.wsj.com/articles/third-covid-shot-vaccine-immunity-israel-study-11628089557

33.   Cohrs, Rachel. "People Chasing Covid-19 Vaccine Boosters Create Headaches for the Healthcare System." *Stat*, August 3, 2021. https://www.statnews.com/2021/08/03/people-chasing-Covid-19-vaccine-boosters-create-headaches-for-the-health-care-system/

34.   Aboulenein, Ahmed. "U.S. FDA clears Pfizer COVID-19 booster dose for older and at-risk Americans." Reuters, September 23, 2021. https://www.reuters.com/world/middle-east/us-fda-authorizes-Covid-19-vaccine-boosters-immunocompromised-2021-08-13/

35.   Lovelace, Berkeley Jr. "Fauci Says Everybody Will Likely Need a Covid Vaccine Booster Shot Eventually." *CNBC*, August 12, 2021. https://www.cnbc.com/2021/08/12/covid-booster-shot-fauci-says-it-is-likely-everybody-will-eventually-need-a-third-vaccine.html

36.   Edwards, Erika. "U.S. Announces Plan to Offer Boosters to All Americans Starting in Late September." *NBC News*, August 18, 2021. https://www.nbcnews.com/health/health-news/u-s-announces-plan-offer-boosters-all-americans-starting-late-n1277059

37.   Lovelace, Berkeley Jr. "The Leader of CDC Just Made a Rare Call to Allow Covid Booster Shots for More People." *CNBC*, September 23, 2021. https://www.cnbc.com/2021/09/23/covid-booster-shots-cdc-panel-endorses-third-Pfizer-doses-for-millions.html

38.   Doria-Rose, Nicole, Mehul S. Suthar, Mat Makowski, et al. "Antibody Persistence through 6 Months after the Second Dose of mRNA-1273 Vaccine for Covid-19." *New England Journal of Medicine* 384 (2021): 2259–61. https://www.nejm.org/doi/full/10.1056/nejmc2103916

39.   Pegu, Amarendra, Sarah O'Connell, Stephen D. Schmidt, et al. "Durability of mRNA-1273-induced antibodies against SARS-CoV-2 variants." [preprint]. BioRxiv, May 16, 2021. https://www.biorxiv.org/content/10.1101/2021.05.13.444010v1

40. Johnson, Carolyn Y. "Yes, We'll All Probably Need a Coronavirus Booster Shot. But Which One?" *Washington Post*, May 27, 2021. https://www.washingtonpost.com/health/2021/05/27/covid-vaccine-booster-shots/

41. Ibid.

42. KHN. "Boosters, Mismatched Doses: Vaccine Innovations May Up Protections." KHN Morning Briefing, May 6, 2021. https://khn.org/morning-breakout/boosters-mismatched-doses-vaccine-innovations-may-up-protections/

43. Johnson, "Yes, We'll All Probably Need a Coronavirus Booster Shot. But Which One?"

44. Mole, Beth. "Health Officials Rail Against Pfizer's Push for COVID Boosters—For Many Reasons." *Ars Technica*, July 12, 2021. https://arstechnica.com/science/2021/07/Pfizer-pushes-for-boosters-as-health-experts-say-theyre-unneeded-unethical/

45. Fidler, Stephen. "WHO Calls for Halt to Covid-19 Booster Shots to Tackle Shortfall in Developing World." *Wall Street Journal*, August 4, 2021. https://www.wsj.com/articles/who-calls-for-halt-to-Covid-19-booster-shots-11628091445

46. "US Rebuffs WHO's Call For Covid Booster Jab Moratorium As China Curbs Travel." Agence France-Presse, August 5, 2021. https://www.ndtv.com/world-news/us-rebuffs-whos-call-for-covid-booster-jab-moratorium-as-china-curbs-travel-2502952

47. Goodenough, Patrick. "US Has Donated the Most Free Vaccines by Far; China Has Sold More Than It's Donated." CNS News, August 4, 2021. https://cnsnews.com/article/international/patrick-goodenough/us-has-donated-most-free-vaccines-far-china-has-sold-more

48. Moderna TX. "Moderna Announces it has Shipped Variant-Specific Vaccine Candidate, mRNA-1273.351, to NIH for Clinical Study." [Press release]. February 24, 2021. https://investors.Modernatx.com/news-releases/news-release-details/Moderna-announces-it-has-shipped-variant-specific-vaccine

49. Moderna TX. "Moderna Announces First Participants Dosed in Study Evaluating COVID-19 Booster Vaccine Candidates." [Press release]. March 10, 2021. https://investors.Modernatx.com/news-releases/news-release-details/Moderna-announces-first-participants-dosed-study-evaluating

50. Author interview with Marcello Damiani, Chief Digital and Operational Excellence Officer at Moderna, April 22, 2021

51. Moderna TX. "Moderna Announces Positive Initial Booster Data Against SARS-CoV-2 Variants of Concern." [Press release]. May 5, 2021. https://investors.Modernatx.com/news-releases/news-release-details/Moderna-announces-positive-initial-booster-data-against-sars-cov

52. Arthur, Rachel. "Moderna to Take New Variant-Specific COVID-19 Vaccine into Phase 1 Trials." *BioPharma Reporter*, February 25, 2021. https://www.biopharma-reporter.com/Article/2021/02/25/Moderna-to-start-Phase-1-study-for-new-variant-specific-COVID-19-vaccine

53. Acharya, Bhargav, and Shilpa Jamkhandikar. "Explainer: What is the Delta variant of coronavirus with K417N mutation?" Reuters, June 23, 2021. https://www.reuters.com/business/healthcare-pharmaceuticals/what-is-delta-variant-coronavirus-with-k417n-mutation-2021-06-23/

54. Hewings-Martin, Yella, and Maria Cohut. "New SARS-CoV-2 Variants: How Can Vaccines Be Adapted?" *Medical News Today*, March 15, 2021. https://www.medicalnewstoday.com/articles/new-SARS-CoV-2-variants-how-can-vaccines-be-adapted

55. Mandavilli, Apoorva, and Benjamin Mueller. "Virus Variants Threaten to Draw Out the Pandemic, Scientists Say." *New York Times*, April 3, 2021. https://www.nytimes.com/2021/04/03/health/coronavirus-variants-vaccines.html

56. Johnson, Carolyn Y. "Coronavirus Vaccines Are Widely Available in the U.S. So Why Are Scientists Working on New Ones?" *Washington Post*, June 30, 2021. https://www.washingtonpost.com/health/2021/06/30/new-coronavirus-vaccines/

57. Hamblin, James. "One Vaccine to Rule Them All." *Atlantic*, April 26, 2021. https://www.theatlantic.com/science/archive/2021/04/finding-universal-coronavirus-vaccine/618701/

58. "3 Questions: Phillip Sharp on the Discoveries that Enabled mRNA Vaccines for Covid-19." *MIT News*. December 11, 2020. https://news.mit.edu/2020/phillip-sharp-rna-vaccines-1211

59. Mandavilli and Mueller, "Virus Variants Threaten to Draw Out the Pandemic, Scientists Say."

60. Ibid.

61. Holpuch, Amanda. "Covid summer: Fauci warns US not to 'declare victory' despite lowest rates in a year." *Guardian*, May 31, 2021. https://www.theguardian.com/us-news/2021/may/31/dr-anthony-fauci-interview-covid-coronavirus-vaccines-summer

62. Taylor, Adam. The Delta Variant Adds a Speed Bump to the Pandemic Escape Route." *Washington Post*, June 17, 2021. https://www.washingtonpost.com/world/2021/06/17/delta-variant-global-concern/

63. Mandavilli and Mueller, "Virus Variants Threaten to Draw Out the Pandemic, Scientists Say."

64. Ibid.

65. Our World in Data. "Coronavirus (COVID-19) Vaccinations." Accessed October 2, 2021. https://ourworldindata.org/covid-vaccinations?country=OWID_WRL

66. WorldOMeter. "COVID-19 Coronavirus Pandemic." Accessed October 2, 2021. https://www.worldometers.info/coronavirus

67. Maxouris, Christina, Lauren Mascarenhas, and Eric Levenson. "Another Coronavirus Surge is Unlikely but the Pandemic Isn't Going Away, Former FDA Chief Says." CNN, March 21, 2021. https://www.cnn.com/2021/03/21/health/us-coronavirus-sunday/index.html

68. Kingsland, James. "COVID-19: Is the B.1.1.7 Variant More Lethal?" *Medical News Today,* March 15, 2021. https://www.medicalnewstoday.com/articles/Covid-19-is-the-b-1-1-7-variant-more-lethal

69. Maxouris, Mascarenhas, and Levenson. "Another Coronavirus Surge Is Unlikely But the Pandemic Isn't Going Away, Former FDA Chief Says."

70. Holpuch, "Covid Summer: Fauci Warns US Not to 'Declare Victory' Despite Lowest Rates in a Year."

71. Stieg, Cory. "How Scientists Can 'Copy and Paste' Covid Vaccines to Work on the Strain from South Africa." *CNBC*, January 28, 2021. https://www.cnbc.com/2021/01/28/why-mrna-vaccines-like-covid-vaccines-are-more-flexible-to-variants.html

72. Tidey, Alice. "COVID and Restrictions Are Here to Stay. Here's Why." *Euronews*, June 11, 2021. https://www.euronews.com/2021/06/11/covid-and-restrictions-are-here-to-stay-here-s-why

73. Domanska, Anna. "IMF says the world needs to commit $50 billion to combat the pandemic. *Industry Leaders.* Accessed October 2, 2021. https://www.industryleadersmagazine.com/imf-says-the-world-needs-to-commit-50-billion-to-combat-the-pandemic/

74. Ibid.

75. Mayo Clinic. "U.S. COVID-19 Vaccine Tracker: See Your State's Progress." Accessed October 2, 2021. https://www.mayoclinic.org/coronavirus-covid-19/vaccine-tracker

76. Murthy, Bhavani Patel, Elizabeth Zell, Ryan Saelee, et al. "COVID-19 Vaccination Coverage Among Adolescents Aged 12–17 Years — United States, December 14, 2020–July 31, 2021." *MMWR Morbidity and Mortality Weekly Report* 70 no. 35 (September 3, 2021):1206–13. doi: http://dx.doi.org/10.15585/mmwr.mm7035e1 https://www.cdc.gov/mmwr/volumes/70/wr/mm7035e1.htm

77. Aschwanden, Christie. "Five reasons why COVID herd immunity is probably impossible." *Nature* 591 (2021): 520-2. https://www.nature.com/articles/d41586-021-00728-2

78. Hinshaw, Drew, and Mike Cherney. "Vaccination Delays Put Global Rebound at Risk." *Wall Street Journal*, January 31, 2021. https://www.wsj.com/articles/vaccination-delays-put-global-rebound-at-risk-11612112184

79. Welsh, Jennifer. "COVID Vaccine Rejectors May Be Here To Stay." Very Well Health, June 25, 2021. https://www.verywellhealth.com/survey-covid-vaccine-rejectors-may-be-here-to-stay-5190366

80. Centers for Disease Control and Prevention. "About Variants." Updated September 20, 2021. https://www.cdc.gov/coronavirus/2019-ncov/variants/variant.html

81. Aschwanden, "Five reasons why COVID herd immunity is probably impossible."

82. Sakay, Yasemin Nicola. "Here's How Well COVID-19 Vaccines Work Against the Delta Variant." Healthline, September 15, 2021. https://www.healthline.com/health-news/heres-how-well-Covid-19-vaccines-work-against-the-delta-variant

83. Georgiou, Aristos. "How Contagious Are Chickenpox, Measles As CDC Document Reveals Delta Variant's R_0." *Newsweek,* July 30, 2021. https://www.newsweek.com/how-contagious-chickenpox-measles-cdc-document-delta-variant-coronavirus-r0-1614661

84. Davies, Scott P., Courtney J. Mycroft-West, Isabel Pagani, et al. "The Hyperlipidaemic Drug Fenofibrate Significantly Reduces Infection by SARS-CoV-2 in Cell Culture Models." *Frontiers in Pharmacology*, August 6, 2021. https://www.frontiersin.org/articles/10.3389/fphar.2021.660490/full

85. Cohan, Alexi. "Cholesterol drug found to reduce coronavirus infection by up to 70%, according to new study." *Boston Herald*, August 14, 2021.

https://www.bostonherald.com/2021/08/14/cholesterol-drug-found-to-reduce-coronavirus-infection-by-up-to-70-according-to-new-study/

86. Merck & Co. Inc. "Merck and Ridgeback's Investigational Oral Antiviral Molnupiravir Reduced the Risk of Hospitalization or Death by Approximately 50 Percent Compared to Placebo for Patients with Mild or Moderate COVID-19 in Positive Interim Analysis of Phase 3 Study." [Press release]. October 1, 2021. https://www.merck.com/news/merck-and-ridgebacks-investigational-oral-antiviral-molnupiravir-reduced-the-risk-of-hospitalization-or-death-by-approximately-50-percent-compared-to-placebo-for-patients-with-mild-or-moderat/

87. Robbins, Rebecca. "Merck Says It Has the First Antiviral Pill Found to Be Effective Against Covid." *New York Times*, October 1, 2021. https://www.nytimes.com/2021/10/01/business/covid-antiviral-pill-merck.html

88. Beasley, Deena. "COVID-19 Pill Developers Aim to Top Merck, Pfizer Efforts." Reuters, September 28, 2021. https://www.reuters.com/business/healthcare-pharmaceuticals/covid-19-pill-developers-aim-top-merck-Pfizer-efforts-2021-09-28/

89. Van Beusekom, Mary. "COVID-19 Most Contagious in First 5 Days of Illness, Study Finds." Center for Infectious Disease Research and Policy, November 20, 2020. https://www.cidrap.umn.edu/news-perspective/2020/11/Covid-19-most-contagious-first-5-days-illness-study-finds

90. Our World in Data. "Coronavirus (COVID-19) Testing." Accessed October 2, 2021. https://ourworldindata.org/coronavirus-testing

91. Chianca, Peter. "A New MIT Study Casts Doubt on the So-Called 6-Foot Rule. Will You Still Keep Your Distance?" Boston.com, April 26, 2021. https://www.boston.com/news/coronavirus/2021/04/26/mit-study-social-distancing-reader-poll/

92. Editorial Board. "Vaccine Mandates Are Working. Let's Make Them the Norm." The Washington Post, September 29, 2021. https://www.washingtonpost.com/opinions/2021/09/29/vaccine-mandates-are-working-lets-make-them-norm/

93. Ibid.

94. Ledford, Heidi. "Six Months of COVID Vaccines: What 1.7 Billion Doses Have Taught Scientists." *Nature*, 594 (2021): 164–7. https://www.nature.com/articles/d41586-021-01505-x

95. Hamblin, "One Vaccine to Rule Them All."

96.   Snyder, Alison, and Bryan Walsh. "Lab Risks Face Scrutiny Amid COVID Origins Controversy." *Axios*, June 10, 2021. https://www.axios.com/ covid-lab-leak-virus-experiment-scrutiny-6d903611-feef-4253-a5ea-98d154c6e356.html

97.   EcoHealth Alliance. "PREDICT Program Info." Accessed October 2, 2021. https://www.ecohealthalliance.org/program/predict

98.   Timmer, John. "Want to Worry About the Next Pandemic? Spillover. global Has You Covered." *Ars Technica,* April 6, 2021. https://arstechnica. com/science/2021/04/want-to-worry-about-the-next-pandemic-spillover-global-has-you-covered/

99.   Buranyi, Stephen. "The mRNA Vaccine Revolution Is Just Beginning." *Wired,* March 6, 2021. https://www.wired.co.uk/article/ mrna-vaccine-revolution-katalin-kariko

100. Oakes, Kari. "Building a Vaccine at Light Speed: mRNA COVID Vaccine Development." Regulatory Affairs Professionals Society, November 4, 2020. https://www.raps.org/news-and-articles/news-articles/2020/11/ building-a-vaccine-at-light-speed-mrna-covid-vacci

101. Labant, MaryAnn. "The Sleeping Giants of Vaccine Production Awaken." Genetic Engineering & Biotechnology News, February 3, 2021. https://www.genengnews.com/topics/drug-discovery/ the-sleeping-giants-of-vaccine-production-awaken/

102. Buranyi, "The mRNA Vaccine Revolution Is Just Beginning."

103. Park, Alice, and Aryn Baker. "Inside the Facilities Making the World's Most Prevalent COVID-19 Vaccine." *Time,* April 19, 2021. https://time. com/5955247/inside-BioNTech-vaccine-facility/

104. AstraZeneca. "Entering a New Era in Vascular and Cardiac Regeneration Research." Accessed October 2, 2021. https://www.AstraZeneca.com/ what-science-can-do/topics/next-generation-therapeutics/entering-a-new-era-in-vascular-and-cardiac-regeneration-research.html

105. Zaks, Tal. "An Important Step in the Advancement of mRNA medicines: Newly Published Clinical Data Show the Early Potential of VEGF-A mRNA as a Regenerative Therapeutic." Moderna Blog, February 20, 2019. https://www.Modernatx.com/Moderna-blog/potential-of-vegf-a-022019

106. Author interview with Robert Langer, Institute Professor at MIT, on March 14, 2021.

107. "siRNA Applications." Horizon Discovery. Accessed October 2, 2021. https://horizondiscovery.com/en/applications/rnai/sirna-applications

108. Author interview with Phillip Sharp, Nobel Prize-winning biochemist and Institute Professor at MIT, on March 17, 2021.

109. "3 Questions: Phillip Sharp on the Discoveries that Enabled mRNA Vaccines for Covid-19."

110. Foley, Katherine Ellen. "The first Covid-19 vaccines have changed biotech forever." *Quartz*, December 22, 2020. https://qz.com/1948132/the-first-Covid-19-vaccines-have-changed-biotech-forever/

111. Cross, Ryan. "Without These Lipid Shells, There Would Be No mRNA Vaccines for COVID-19." *Chemical & Engineering News*, March 6, 2021. https://cen.acs.org/pharmaceuticals/drug-delivery/Without-lipid-shells-mRNA-vaccines/99/i8

112. Foley, "The first Covid-19 vaccines have changed biotech forever."

113. *Washington Post*, February 7, 2021. https://www.washingtonpost.com/lifestyle/magazine/Pfizer-ceo-on-the-pressures-of-creating-a-Covid-19-vaccine-what-is-at-stake-is-beyond-imagination/2020/09/29/2ddbe7f8-fdb3-11ea-b555-4d71a9254f4b_story.html

114. Buranyi, "The mRNA Vaccine Revolution Is Just Beginning."

115. Ustinova, Anastasia. "In the Thick of the 'Herculean' Vaccine Push." SME Media, September 21, 2020. https://www.sme.org/technologies/articles/2020/september/vaccine-placeholder/

116. Eastman, Peggy. "Oncologists Still Call FDA Too Slow to Approve Drugs but Opinion Better than in 1995 Poll." *Oncology Times*, 24 no. 7 (July 2002): 1, 65-6. https://journals.lww.com/oncology-times/fulltext/2002/07000/Oncologists_Still_Call_FDA_Too_Slow_to_Approve.3.aspx

117. Gibson, Beth, Daniel J. Wilson, Edward Feil, and Adam Eyre-Walker. "The distribution of bacterial doubling times in the wild." *Proceedings of the Royal Society of London. Series B, Biological Sciences* 285 no. 1880 (2018): 20180789. https://www.ncbi.nlm.nih.gov/pmc/articles/PMC6015860/

118. SAFC. "Extended, High Density Growth of CHO-K1 Cells in EX-CELL® 302 Serum-Free Medium. [Brochure]. Accessed October 2, 2021. https://www.sigmaaldrich.com/content/dam/sigma-aldrich/docs/Sigma/General_Information/2/r003.pdf

119. Fleshler, David. "The Gulf Stream Is Slowing Down. That Could Mean Rising Seas and a Hotter Florida." Phys.org, August 9, 2019. https://phys.org/news/2019-08-gulf-stream-seas-hotter-florida.html

120. Roser, Max. "Why Did Renewables Become So Cheap So Fast?" Our World In Data, December 1, 2020. https://ourworldindata.org/cheap-renewables-growth

121. IW Staff. "Next Gen Batteries to Power Up Electric Vehicle Installed Base to 100 Million by 2028." *Industry Week*, August 9, 2019. https://www.industryweek.com/technology-and-iiot/article/22028055/next-gen-batteries-to-power-up-electric-vehicle-installed-base-to-100-million-by-2028

122. "Energy Storage Grand Challenge Roadmap." U.S. Department of Energy. December 2020. Accessed October 2, 2021. https://www.energy.gov/sites/prod/files/2020/12/f81/Energy%20Storage%20Grand%20Challenge%20Roadmap.pdf

123. Naujokaitytė, Goda. "Number of Scientists Worldwide Reaches 8.8M, as Global Research Spending Grows Faster than the Economy." *Science | Business*, June 14, 2021. https://sciencebusiness.net/news/number-scientists-worldwide-reaches-88m-global-research-spending-grows-faster-economy

Index

A

adenovirus 18, 67, 86
Africa 25, 63, 75, 97, 107, 114, 121
AIDS 75, 97. *See also* HIV/AIDS
Amazon 47, 76
amino acids 4
antibodies 3, 10–14, 107–109,
    121–122
antigen viii, 3–10, 18, 29–31
    spike protein viii, 3–10, 29–31,
    104, 106
anti-vaccine sentiment 81,
    90–95, 100–102, 114, 130
    in Europe 92
Apollo space program vii–viii
Aquino, Benigno, III.
    See Philippines: government of
artificial intelligence 10
AstraZeneca 20, 64, 86, 122
AstraZeneca vaccine. *See* Oxford–
    AstraZeneca vaccine
Australia 106

B

bacteria 3, 120
Bahrain 68
Bancel, Stéphane. *See* Moderna
Bar-Ilan University 50, 76
    Epstein, Gil 50–52, 76
basic reproduction num-
    ber 1, 71–72, 106, 115–118. *See*
    also epidemiology
Bharat Biotech 53, 66

Covaxin 66
Biden, Joe 60, 68, 80, 97–98, 109.
    See also United States: govern-
    ment of
    administration of 68, 97
biochemistry 6–7, 24, 121
biology
    cell biology viii, 120
    molecular biology 4, 43
BioNTech xi, 6, 21, 35, 39, 42, 107,
    121
    Covid-19 vaccine. *See* Pfizer-
    BioNTech vaccine
    Karikó, Katalin 5, 6
    Kuhn, Andreas 121
    Şahin, Uğur 8
biopharmaceuticals 10
biotechnology 6–9, 21, 32, 39,
    65–67, 124
Bolsonaro, Jair. *See* Brazil: gov-
    ernment of
Bourla, Albert. *See* Pfizer
Brazil 15, 67–68, 96
    government of
    Bolsonaro, Jair 96
bubonic plague 1
Bush, George W. 97. *See*
    also United States: government
    of

C

Canada 42, 55–56, 68, 126
  government of 55–56
  Trudeau, Justin 56
cancer ix, 65, 122–124
CanSino 17
CDC (Centers for Disease Control
  and Prevention) 74, 83–89,
  104–108. *See also* United States:
  government of: federal agencies
  Walensky, Rochelle 104, 108
cell biology viii, 120
chemical engineering 24
chemistry 24, 29, 33, 36, 127
China 9–10, 17, 28, 65–68, 77,
  96–99, 125
  Chinese-made vaccines 17,
    65–68
  government of 28, 98
    Chinese Communist
      Party 99
  Wuhan 28, 72, 96–97
clinical trials 5, 11–16, 32, 40, 49,
  67, 82, 91, 109, 116–117, 122
  candidate vaccines 2, 5, 8,
    11–15, 19–20, 61, 123
  challenge trials 14
  double-blind protocol 13–16,
    93, 96
  ethics 11–15
    1796 cowpox vaccine trial 11
    Henrietta Lacks cell
      studies 88
    1932–72 Tuskegee syphilis
      study 88
  in silico 12
  in vitro 11, 41, 116
  in vivo 12, 41
  Phase 1 13, 17, 59
  Phase 2 13, 16, 136
  Phase 3 13–18, 45, 49, 61, 67, 116
Coalition for Epidemic Preparedness Innovations 20

cold-chain handling 43–45, 48,
  57, 149–150
confirmation bias 90
contact tracing 10, 113, 118. *See
  also* pandemic: control measures
CordenPharma 33, 144
CoronaVac 67–68
coronavirus 9–10, 49, 95–98, 106,
  110, 121
COVAX 20, 68
Covaxin 66
Covid-19 vaccines
  CoronaVac 11
  Covaxin 66
  Covishield. *See* Oxford–Astra-
    Zeneca vaccine
  Johnson & Johnson 23, 66,
    82–86
  Moderna viii, 8, 11, 13–18, 31,
    33, 43, 56, 116, 121, 123
  Novavax 20, 66
  Oxford–AstraZeneca 8, 17–18,
    66, 68, 85, 107
  Pfizer-BioNTech viii, 8, 11,
    15–21, 26, 37, 43–49, 52, 56, 68,
    85, 97–98, 107, 116, 123
  Sputnik V 8, 17, 66–67, 86
Covishield. *See* Oxford–AstraZen-
  eca vaccine
Crick, Francis. *See* DNA
cruise lines 71, 81
  Carnival 81
  Norwegian 81
  Virgin 81
CureVac 6–7, 21, 110

D

Damiani, Marcello. *See* Moderna
data 9–17, 40–45, 50, 55–56, 61,
  67, 82–86, 90–97, 105–109, 113,
  115, 129–130
Dengvaxia 87
deoxyribonucleic acid. *See* DNA

W

Walensky, Rochelle. *See* CDC
(Centers for Disease Control and
Prevention)
Wall Street 80
Wall Street Journal 26, 40, 95
Watson, James. *See* DNA
Weissman, Drew. *See* Pennsylva-
nia, University of
Western medicine 75–77, 97–98.
See also traditional medicine
WHO (World Health Organiza-
tion) 10, 20, 51, 95, 105, 109, 111

Z

ZEDpharma 26
Zika 121

Printed in Great Britain
by Amazon